# The River of the West, The Mountain Years

# Also by Win Blevins

*Fitz*

*Rendezvous: Volume One*

*Rendezvous: Volume Two*

*Rendezvous: Volume Three*

**Rivers of the West**

*The Yellowstone*

*The High Missouri*

*The Snake River*

*The Powder River*

# The River of the West, The Mountain Years

## The Adventures of Joe Meek Part One

### Classics of the Fur Trade

### Book 1

**Win Blevins**

**The River of the West, The Mountain Years: The Adventures of Joe Meek, Part One**
Paperback Edition


Wolfpack Publishing
701 S. Howard Ave. 106-324
Tampa, Florida 33609

wolfpackpublishing.com



Paperback ISBN 978-1-63977-520-0
eBook ISBN 978-1-63977-519-4

*We find them, accordingly, hardy, lithe, vigorous, and active. Extravagant in word, in thought, and deed. Heedless of hardship, daring of danger, prodigal of the present, and thoughtless of the future.*

**— W. Irving**

# Introduction

## The Person and Persona of Joe Meek

Come meet Joe Meek. This is his book, the tale of his doings in the Rocky Mountains. He tells it in high style. The Joe he yarns up is one of the West's irresistible characters—dashing, devil-may-care, cheeky, irreverent, more fun than a playful grizzly cub. He will play every sort of prank on his companions and maybe on you too. He likes a good story and has enough sense to tell it tall. He has the spirit to spin some stories at his own expense, but the grace to tell none that demean a companion. He tells no sad tales and sits still for none: "Sympathy is repudiated by your true mountaineer for himself, nor will he furnish it to others. The absurd and humorous, or the daring and reckless, side of a story is the only one which is dwelled upon in narrating his adventures. The laugh which is raised at his expense when he has a tale of woes to communicate is a better tonic to his dejected spirits than the gentlest pity would be. Thus lashed into courage again, he is ready to declare that all his troubles were only so much pastime."

That's Joe, full of thumb-your-nose-at-the-universe sass. His recollections are a barbaric yawp of joy—joy in life, joy in youth and vitality, exultation at challenge and danger. Of all the men we can find buried in the records of those who walked the Rocky Mountains a century and a half ago, none other is so alive.

So, is the Joe of these pages the real Joe? Or was he the sort who lived it one way and told it another?

This question is always tricky, but so far as after-the-fact tests can tell, Joe seems resoundingly the man of his words.

The large picture of his character and spirit seems true enough. The woman who here has written his tale down for him, Frances Fuller Victor, described the 58-year-old man she saw as "a tall, broad-shouldered, powerful and handsome man, with plenty of animal courage and spirit, though somewhat at the expense of the inner furnishing which is supposed to be necessary to a perfect development. In this instance, however, nature had been more than usually kind and distinguished her favorite with a sort of inborn grace and courtesy which...served him well." His fellow Oregonians thought him reliable, right-minded, and rough enough to be their first sheriff and first marshal. Even his critics confirm our winning portrait of Joe if we do not share all their values: Joe offended the missionary contingent in Oregon with his irreverence, and they complained that he was too free-handed, that he preferred hunting and fighting to manual labor, and that he wouldn't give up teasing them when they tried to convert him.

It's just as well that they didn't know how Joe got his wife Virginia, whom the Oregon whites snubbed because she was

Nez Percé. Probably, the missionaries and their wives feared that Joe had bought her, or traded for her, or made some such barbaric transaction. If only they had known what he traded for Virginia:

In 1838, settling in for the winter with the Nez Percé, Joe found himself temporarily womanless—his last wife, a Nez Percé, had run off on him. And by chance, Meek found that the Indians were keen to hear preaching. They'd been hearing tales of the marvels taught by the missionaries on the Walla-Walla, and they wanted to learn such wonders for themselves. Seeing an opportunity, Meek decided he recalled enough theology to plug the breech. He carefully swore his fellow trappers to secrecy, volunteered his services, and proceeded to hold forth intermittently for several weeks. What he accomplished, other than to amuse his friends, is best left uncontemplated. At Christmas, Joe decided to let the chief know what would be an appropriate gift to reward the dispensing of the gospel, according to Meek—a wife. And so, after some discussion, he acquired Virginia without even having to give up any horseflesh.

The rest of the story—crucial to Joe's character—is that when he left the mountains a year and a half later, he took Virginia with him to the white settlements, where he knew she would be belittled as a squaw and he as a squawman. They formalized the marriage and lived together until his death thirty-five years later. She bore him seven children—so Joe's loyalty is ballast to his levity.

So what about the small picture, the anecdotes that Joe tells? Some of Joe's contemporaries in Oregon pooh-poohed them as mere tall tales. Mrs. Victor said simply—both in her introduction and in a later magazine article[1]—that, not doubting the stories as Joe gave them to her, she was passing them on intact. And it's true, some of them are surely stretchers, which may disturb some folks who don't realize that a true story yarned big is not a lie. I do not think that Joe can be convicted of having told any lies, and one of my reasons is that he gets so much corroboration from his comrades. The notes at the back of his book show how often one of Joe's anecdotes has also been told by someone else on hand—Osborne Russell or Kit Carson or Jim Beckwourth or Captain William Drummond Stewart or Captain Benjamin Bonneville or Joe's brother Stephen or his lifelong friend Doc Newell and they're consistently in versions consonant with Joe's.

Most important, maybe, not a single alternative version of one of Joe's tales makes him look like he was covering up, trying to justify himself, being vindictive, acting petty, or committing any such sins that autobiographers are usually prey to. On the contrary, his stories are often humorously self-deprecating. He seems sensible, pragmatic, democratic, fair, and intolerant of humbug in others and in himself. Such a man commands confidence.

Accepting Joe's mountain portrait, in essence, does not mean believing all the details. Joe does not stand above the time-honored Western custom of shucking dudes. I would not care, for instance, to subscribe publicly to the notion that Joe, at his first rendezvous, saw trappers using the body of a dead comrade as a card table. But be warned: Anyone who objects to thus and

1. Introduction, v, *Pacific Christian Advocate*, June 11, 1870, p. 3—quoted in Hazel E. Mills, "The Emergence of Frances Fuller Victor—Historian," *Oregon Historical Quarterly*, Vol. LXII, No. 4, December, 1961, p. 321.

such a story as improbable and even outrageous is likely to get caught out. Common observation and historical research alike show that life is nothing if not improbable and outrageous. And Joe Meek had a yen to be both.

If one principle could be genetically implanted into future historians, it should be: Don't figure out what they must have done, or what is reasonable to assume they did, or what logic says they did—find out what they actually did.

About Joe's chronology: Joe had no memory for dates. He could mimic a fellow trapper's voice years later, but he couldn't always remember what winter it was the fellow told that tale. In recollecting, Joe was trying to ford a stream thirty years wide and deep. Better to rely on Osborne Russell for dates—he kept a diary.

Russell is, in general, more useful for facts. He doesn't show Joe's inclination to exaggerate the number of Indians killed or horses stolen. He's scrupulous, detailed, reliable—fact-oriented. Joe isn't always fastidious about details, but he has a virtue Russell lacks—he catches the spirit of the thing. I am glad Russell's *Journal of a Trapper* provides the particulars, and more glad Joe's *River of the West* gives us the vibrant truth.

Not that Joe is a paragon. We may excuse what he describes as the *meanest act of his life*—snitching the negro boy's dinner—as hijinks. We can only hope that his shooting a Digger Indian because the fellow *looked as if he was going to* steal some traps is one of Joe's exaggerations. And we can wonder why he made an entire book without once mentioning a frequent trapping companion who was also his older brother, Stephen Hall Meek. Yet, on the whole he seems to meet Gary Cooper's worthy definition of a hero—not a knight in shining armor but just a right kind of fella.

In his essential innocence and optimism, his vitality, his confidence in himself, and his zest for life, Joe reminds me of a matured Huckleberry Finn. I have a fantasy that Joe is Huck, grown tall without losing all his boyishness. Why not? At the end of *The Adventures of Huckleberry Finn*, Huck resolves to get away from Aunt Polly and civilization and all such confining things and go where a fellow can be free and have some fun. Gonna light out for the territories, he says. Like Joe did. Joe trod the mountains freely, welcomed challenges, soaked up the good times, and kept his boyish enthusiasm through everything—like Huck would have. Neither liked the Aunt Pollys of the world with their moralizing and their burdensome seriousness. Life was too much fun for that.

---

## The Creation of the River of the West

It is one of the serendipities of the writing of Western history that Joe Meek met Frances Fuller Victor. She was a professional writer—a poet, a magazine columnist, and a would-be historian. Following her second husband to Oregon in 1864, she started gathering material for a book on the history of her new state. Soon, she was introduced to *Uncle Joe*, as he was familiarly known. Folks had been telling Joe to set his memories down in a book. Off the two of them set on a literary adventure.[2]

---

2. Biographical details about Mrs. Victor are taken from *The Reader's Encyclo-*

Joe would tell stories. Mrs. Victor would question and cross-question, drawing Joe out. She was particularly concerned to get his colorful language down, but despaired of capturing his wonderful mimicry on paper. When they weren't having long interviews, Joe would send notes over from his home, notes made by his daughter Olive. Mrs. Victor questioned other old-time Oregonians about the history of the Territory and State: the members of the missionary contingent, the first to arrive, the son-in-law of the former governor of Fort Vancouver, people prominent in territorial politics, Judge Matthew Deady, who gathered printed materials for her. She read what documents she could get her hands on.[3] And at length she brought forth her book, mostly Joe's biography, with long threads of Oregon history woven in here and there. R. W. Bliss published it in Hartford, Connecticut, in 1870.

---

## Publication & Reception

The Methodists did not like it. They thought her prejudiced against them and their role in the development of Oregon. Earlier on, the missionaries and the small colony of trappers had been opposing cliques. Some people wanted to dismiss Joe's

---

*pedia of the American West*, edited by Howard Lamar—New York, 1977—p. 1225-26.

3. This account of the composition *of River of the West* is taken from Mrs. Victor's introduction to that book, vii, Mrs. Mills's article cited above in note 3, and from a letter from Mrs. Mills to me dated March 29, 1983.

stories as trapper rubbish. But most people liked the book. It was reprinted in the year of its publication and again the next year.

In 1877, after Joe's death, Mrs. Victor brought the book out again, in revised form, under a new *title, Eleven Years in the Rocky Mountains*. Here, she made it more truly a biography of Meek, deleting her material on Oregon history that did not involve Meek personally, including nearly the last 100 pages, correcting some errors of fact, and eliminating some of what offended the missionary contingent. Unfortunately, she also added a section on George Armstrong Custer and the slaughter on the Little Big Horn, doubtless to boost sales.[4]

Though popular, *River of the West* and *Eleven Years in the Rocky Mountains* never sold well enough to set Mrs. Victor's finances right. She had been in financial difficulty since her separation from her husband in 1868. She continued to write prolifically to earn a living, composing the histories of Oregon, Washington, Idaho, Montana, Nevada, Colorado, and Wyoming that Hubert Howe Bancroft published under his own name. She worked actively until her death in 1902 at the age of 76.

Though the book has been mostly recognized simply as Joe Meek's life story, and Mrs. Victor recognized simply as a good conduit, she did add at least one perception of her own that merits historical recognition: She must have been one of the first white writers to understand that the intense desire of the Northwest tribes to know the Christian God had to do not at all with the spiritual realm but with the material. They understood that

4. This bibliographical information on *River of the West* and *Eleven Years in the Rocky Mountains* comes from the letter from Hazel Mills cited in note 7. I have not been able to see *Eleven Years in the Rocky Mountain* myself.

prayers and chants brought them material things, like buffalo, for food, clothing, and shelter. They saw that the whites had watches and magnifying glasses and guns and other goods their gods did not bring them. They wanted to learn enough about the white God to get such things. To change their way of living, to grow from so-called benightedness to so-called enlightenment—these never occurred to the Indians. Yet they were foremost in the minds of the missionaries. There germinated a tragic misunderstanding, and Mrs. Victor names it squarely.

As Joe's record, the book was read and admired at the time and has commanded the respect of historians since. Hiram Chittenden used it as a source for his major work on the mountain fur trade, *The History of the American Fur Trade in the Far West*, and wrote to Mrs. Victor that her book "will always stand, not only as a pioneer, but as one of the ablest examples in historical work of the fur trading era."[5] J. Cecil Alter, Jim Bridger's biographer, in his preface to the reprint of *River of the West* by Long's College Book Company in 1950—Columbus, Ohio—calls Joe's book "easily the most reliable and extensive first-person history ever written of that unique and never-to-be duplicated industry and era."

5. Chittenden, New York, 1902. Comment from a letter to Mrs. Victor, quoted by her to Olive C. Applegate in a letter dated September 16, 1896. I have not seen the letter myself, but quoted from Mills, the article cited above, p. 311.

# Author's Note

When absorbed in the elegant narratives of Washington Irving, reading and musing over *Astoria* and *Bonneville*, in the cozy quiet of a New York study, I had no prescient motion of the mind ever gave prophetic indication of that personal acquaintance which has since been formed with the scenes, and even with some of the characters which figure in the works just referred to. Yet so have events shaped themselves that to me Astoria is familiar ground. Forts Vancouver and Walla-Walla pictured forever in my memory, while such journeys as I have been enabled to make into the country east of the last-named fort, have given me a fair insight into the characteristic features of its mountains and its plains.

Today, a railroad traverses the level stretch between the Missouri River and the Rocky Mountains, along which, thirty years ago, the fur traders had worn a trail by their annual excursions with men, packhorses, and sometimes wagons, destined to the Rocky Mountains. Then, they had to guard against the attacks of the Savages, and in this respect, civilization is behind the railroad, for now, as then, it is not safe to travel without a sufficient escort. Today, also, we have new Territories called by several names cut out of the identical hunting grounds of the fur traders of thirty years ago and steamboats plying the rivers where the mountain men came to set their traps for beaver or cities growing up like mushrooms from a soil made quick by gold, where the hardy mountain hunter pursued the buffalo herds in search of his winter's supply of food.

The wonderful romance which once gave enchantment to stories of hardship and of daring deeds, suffered and done in these then distant wilds, is fast being dissipated by the rapid settlement of the new Territories, and by the familiarity of the public mind with tales of stirring adventure encountered in the search for glittering ores. It was, then, not without an emotion of pleased surprise that I first encountered in the fertile plains of Western Oregon the subject of this biography, a man fifty-eight years of age, of fine appearance and buoyant temper, full of anecdote, and with a memory well stored with personal recollections of all the men of note who have formerly visited the old Oregon Territory, when it comprised the whole country west of the Rocky Mountains lying north of California and south of the forty-ninth parallel. This man is *Joseph L. Meek*, to whose stories of mountain life I have listened for days together, and who, after having figured conspicuously, and not without considerable fame, in the early history of Oregon, still prides himself most of all on having been a *mountain man*.

Most persons are familiar with the popular, celebrated Indian pictures of the artist Stanley, and it cannot fail to interest the reader to learn that in one of these, Meek is represented as firing his last shot at the pursuing Savages. He was also the hero of another picture, painted by an English artist. The latter picture represents him in a contest with a grizzly bear, and has been copied in wax for the benefit of a St. Louis Museum, where it has been repeatedly recognized by Western men.

It has frequently been suggested to Mr. Meek, who has now come to be known by the familiar title of *Uncle Joe* to all Oregon, that a history of his varied adventures would make a readable book, and some of his neighbors have even undertaken to become his historian, yet with so little well-directed efforts that the task, after all, has fallen to a comparative stranger. I confess to having taken hold of it with some doubts as to my claims to the office, and the best recommendation I can give my work is the interest I myself felt in the subject of it, and the only apology I can offer for anything incredible in the narrative which it may contain, is that I *tell the tale as 'twas told to me*, and that I have no occasion to doubt the truth of it.

Mr. Meek has not attempted to disguise the fact that he, as a mountain man, *did those things which he ought not to have done, and left undone those things which he ought to have done.* It will be seen by referring to Mr. Irving's account of this class of men, as given to him by Capt. Bonneville, that he in no wise differed from the majority of them in his practical rendering of the moral code, and his indifference to some of the commandments. Yet, no one seeing Uncle Joe in his present aspect of a good-humored, quiet, and not undignified citizen of the *Plains*, would be likely to attribute to him any very bad or dangerous qualities. It is only when recalling the scenes of his early exploits in mountain life, that the smoldering fire of his still fine eyes brightens up with something suggestive of the dare-devil spirit which characterized those exploits, and made him famous even among his compeers, when they were such men as Kit Carson, Peg-Leg Smith, and others of that doughty band of bear-fighters.

Seeing that the incidents I had to record embraced a period of a score and a half of years, and that they extended over those years most interesting in Oregon history, as well as of the history of the Fur Trade in the West, I have concluded to preface Mr. Meek's adventures with a sketch of the latter, believing that the information thus conveyed to the reader will give an additional degree of interest to their narration. The impression made upon my own mind as I gained a knowledge of the facts which I shall record in this book relating to the early occupation of Oregon, was that they were not only profoundly romantic, but decidedly unique.

In giving Mr. Meek's personal adventures, I should have preferred always to have clothed them in his own peculiar language could my memory have served me, and above all, I should have wished to convey to the reader some impression of the tones of his voice, both rich and soft, and deep, too—or suddenly changing, with a versatile power quite remarkable, as he gave with natural dramatic ability the perfect imitation of another's voice and manner. But these fine touches of narrative are beyond the author's skill, and the reader must perforce be content with words, aided only by his own powers of imagination in conjuring up such tones and subtle inflections of voice as seem to him to suit the subject. Mr. Meek's pronunciation is Southern. He says *thar*, and *whar*, and *bar*, like a true Virginian as he is, being a blood relation of one of our Presidents from that State, as well as cousin to other one-time inmates of the White

House. Like the children of many other slave-holding planters, he received little attention and was allowed to frequent the negro quarters while the alphabet was neglected. At the age of sixteen he could not read. He had been sent to a school in the neighborhood, where he had the alphabet set for him on a wooden *paddle*, but not liking this method of instruction, he one day *hit the teacher over the head with it, and ran home*, where he was suffered to disport himself among his Black associates, clad like themselves in a tow frock, and guiltless of shoes and stockings. This sort of training was not without its advantages to the physical man. On the contrary, it produced, in this instance, as in many others, a tall, broad-shouldered, powerful and handsome man, with plenty of animal courage and spirit, though somewhat at the expense of the inner furnishing which is supposed to be necessary to a perfect development. In this instance, however, nature had been more than usually kind, and distinguished her favorite with a sort of inborn grace and courtesy which, in some phases of his eventful life, served him well.

Mr. Meek was born in Washington Co., Virginia, in 1810, one year before the settlement of *Astoria*, and at a period when Congress was much interested in the question of our Western possessions and their boundary. *Manifest destiny* seemed to have raised him up, together with many others, bold, hardy, and fearless men, to become sentinels on the outposts of civilization, securing to the United States with comparative ease a vast extent of territory, for which, without them, a long struggle with England would have taken place, delaying the settlement of the Pacific Coast for many years, if not losing it to us altogether. It is not without a feeling of genuine self-congratulation, that I am able to bear testimony to the services, hitherto hardly recognized, of the *mountain men* who have settled in Oregon. Whenever there shall arise a studious and faithful historian, their names shall not be excluded from honorable mention, nor least illustrious will appear that of Joseph L. Meek, the Rocky Mountain Hunter and Trapper.

# Preface

[1]In the year 1818, Mr. Prevost, acting for the United States, received Astoria back from the British, who had taken possession, as narrated by Mr. Irving, four years previous. The restoration took place in conformity with the treaty of Ghent, by which those places captured during the war were restored to their original possessors. Mr. Astor stood ready at that time to renew his enterprise on the Columbia River, had Congress been disposed to grant him the necessary protection which the undertaking required. Failing to secure this, when the United States sloop of war Ontario sailed away from Astoria, after having taken formal

1. The particulars of Mrs. Victor's history of the mountain fur trade are not to be relied on. She seems to have gleaned sounder information on the history of Hudson's Bay Company in Oregon in the early years than on the doings of the American trappers. Doubtless she had better authorities and documents for the Hudson's Bay material. If she depended heavily on Meek—other American mountain men were within her reach—it should be remembered that Joe came to the mountains after most of these events were over, and that he seems to have had no knack for dates.

possession of that place for our government, the country was left to the occupancy—scarcely a joint occupancy, since there were then no Americans here—of the British traders. After the war, and while negotiations were going on between Great Britain and the United States, the fort at Astoria had remained in possession of the North-West Company, as their principal establishment west of the mountains. It had been considerably enlarged since it had come into their possession and was furnished with artillery enough to have frightened into friendship a much more warlike people than the subjects of old king Comcomly, who, it will be remembered, was not at first very well disposed toward the *King George men*, having learned to look upon the *Boston men* as his friends in his earliest intercourse with the whites. At this time, Astoria, or *Fort George*, as the British traders called it, contained sixty-five inmates, twenty-three of whom were whites, and the remainder Canadian half-breeds and Sandwich Islanders. Besides this number of men, there were a few women, the native wives of the men, and their half-breed offspring. The situation of Astoria, however, was not favorable, being near the seacoast, and not surrounded with good farming lands such as were required for the furnishing of provisions to the fort. Therefore, when in 1821 it was destroyed by fire, it was only in part rebuilt, but a better and more convenient location for the headquarters of the North-West Company was sought for in the interior.

About this time, a quarrel of long-standing between the Hudson's Bay and North-West Companies culminated in a battle between their men in the Red River country, resulting in a considerable loss of life and property. This affair drew the attention of the government at home—the rights of the rival companies were examined into, the mediation of the Ministry secured, and a compromise effected, by which the North-West Company, which had succeeded in dispossessing the Pacific Fur Company under Mr. Astor, was merged into the Hudson's Bay Company, whose name and fame are so familiar to all the early settlers of Oregon.

At the same time, Parliament passed an act by which the hands of the consolidated company were much strengthened, and the peace and security of all persons greatly insured, but which became subsequently, in the joint occupancy of the country, a cause of offense to the American citizens, as we shall see hereafter. This act allowed the commissioning of Justices of the Peace in all the territories not belonging to the United States, nor already subject to grants. These justices were to execute and enforce the laws and decisions of the courts of Upper Canada, to take evidence, and commit and send to Canada for trial the guilty, and even in some cases, to hold courts themselves for the trial of criminal offenses and misdemeanors not punishable with death, or of civil causes in which the amount at issue should not exceed two hundred pounds.

Thus, in 1824, the North-West Company, whose perfidy had occasioned such loss and mortification to the enterprising New York merchant, became itself a thing of the past, and a new rule began in the region west of the Rocky Mountains. The old fort at Astoria having been only so far rebuilt as to answer the needs of the hour, after due consideration, a site for headquarters was selected about one hundred miles from the sea, near the mouth of the Willamette River, though opposite to it. Three considerations went to make up the eligibility of the point selected. First, it was desirable, even necessary, to settle upon good agricultural lands, where the Company's provisions could be raised by the

Company's servants. Second, it was important that the spot chosen should be upon waters navigable for the Company's vessels, or upon tidewater. Last, and not least, the Company had an eye to the boundary question between Great Britain and the United States, and believing that the end of the controversy would probably be to make the Columbia River the northern limit of the United States territory, a spot on the northern bank of that river was considered a good point for their fort, and possible future city.

The site chosen by the North-West Company in 1821, for their new fort, combined all these advantages, and the further one of having been already commenced and named. Fort Vancouver became at once on the accession of the Hudson's Bay Company, the metropolis of the northwest coast, the center of the fur trade, and the seat of government for that immense territory, over which roamed the hunters and trappers in the employ of that powerful corporation. This post was situated on the edge of a beautiful sloping plain on the northern bank of the Columbia, about six miles above the upper mouth of the Wallamet. At this point, the Columbia spreads to a great width, and is divided on the south side into bayous by long sandy islands, covered with oak, ash, and cottonwood trees, making the noble river more attractive still by adding the charm of curiosity concerning its actual breadth to its natural and ordinary magnificence. Back of the fort, the land rose gently, covered with forests of fir, and away to the east swelled the foothills of the Cascade Range, then the mountains themselves, draped in filmy azure and over-topped five thousand feet by the snowy cone of Mt. Hood.

In this lonely situation grew up, with the dispatch which

characterized the acts of the Company, a fort in most respects similar to the original one at Astoria. It was not, however, thought necessary to make so great a display of artillery as had served to keep in order the subjects of Comcomly. A stockade enclosed a space about eight hundred feet long by five hundred broad, having a bastion at one corner, where were mounted three guns, while two eighteen pounders and two swivels were planted in front of the residence of the governor and chief factors. These commanded the main entrance to the fort, besides which there were two other gates in front, and another in the rear. Military precision was observed in the precautions taken against surprises, as well as in all the rules of the place. The gates were opened and closed at certain hours, and were always guarded. No large number of Indians were permitted within the enclosure at the same time, and every employee at the fort knew and performed his duty with punctuality.

The buildings within the stockade were the governor's and chief factors' residences, stores, offices, workshops, magazines, warehouses, etc.

Year by year, up to 1835 or '40, improvements continued to go on in and about the fort, the chief of which was the cultivation of the large farm and garden outside the enclosure, and the erection of a hospital building, large barns, servants' houses, and a boat-house, all outside of the fort—so that at the period when the Columbia River was a romance and a mystery to the people of the United States, quite a flourishing and beautiful village adorned its northern shore, and that too erected and sustained by the enemies of American enterprise on soil commonly believed to belong to the United States: fair foes the author firmly believes them to have been in those days, yet foes nevertheless.

The system on which the Hudson's Bay Company conducted its business was the result of long experience, and was admirable for its method and its justice also. When a young man entered its service as a clerk, his wages were small for several years, increasing only as his ability and good conduct entitled him to advancement. When his salary had reached one hundred pounds sterling, he became eligible for a chief-tradership as a partner in the concern, from which position he was promoted to the rank of a chief factor. No important business was ever entrusted to an inexperienced person, a policy which almost certainly prevented any serious errors. A regular tariff was established on the Company's goods, comprising all the articles used in their trade with the Indians, nor was the quality of their goods ever allowed to deteriorate. A price was also fixed upon furs according to their market value, and an Indian knowing this, knew exactly what he could purchase. No bartering was allowed. When skins were offered for sale at the fort, they were handed to the clerk through a window like a post-office delivery-window, and their value in the article desired, returned through the same aperture. All these regulations were of the highest importance to the good order, safety, and profit of the Company. The confidence of the Indians was sure to be gained by the constancy and good faith always observed toward them, and the Company obtained thereby numerous and powerful allies in nearly all the tribes.

As soon as it was possible to make the change, the Indians were denied the use of intoxicating drinks, the appetite for which had early been introduced among them by coasting vessels, and even continued by the Pacific Fur Company at Astoria. It would have been dangerous to have suddenly deprived them of the coveted stimulus, therefore, the practice must be discontinued by many wise arts and devices. A public notice was given that the sale of it would be stopped, and the reasons for this prohibition explained to the Indians. Still, not to come into direct conflict with their appetites, a little was sold to the chiefs, now and then, by the clerks, who affected to be running the greatest risks in violating the order of the company. The strictest secrecy was enjoined on the lucky chief who, by the friendship of some under-clerk, was enabled to smuggle off a bottle under his blanket. But the cunning clerk had generally managed to get his *good friend* into a state so cleverly between drunk and sober, before he entrusted him with the precious bottle, that he was sure to betray himself. Leaving the shop with a mien even more erect than usual, with a gait affected in its majesty, and his blanket tightened around him to conceal his secret treasure, the chuckling chief would start to cross the grounds within the fort. If he was a new customer, he was once or twice permitted to play his little game with the obliging clerk whose particular friend he was, and to escape detection.

But by-and-by, when the officers had seen the offense repeated more than once from their purposely contrived posts of observation, one of them would skillfully chance to intercept the guilty chief at whose comical endeavors to appear sober he was inwardly laughing, and charge him with being intoxicated. Wresting away the tightened blanket, the bottle appeared as evidence that could not be controverted, of the duplicity of the Indian and the unfaithfulness of the clerk, whose name was instantly demanded, that he might be properly punished. When the chief again visited the fort, his particular friend met him with a sorrowful countenance, reproaching him for having been the cause of his disgrace and loss. This reproach was the surest means of preventing another demand for rum, the Indian being too magnanimous, probably, to wish to get his friend into trouble, while the clerk affected to fear the consequences too much to be induced to take the risk another time. Thus, by kind and careful means, the traffic in liquors was at length broken up, which otherwise would have ruined both Indian and trader.

To the company's servants, liquor was sold or allowed at certain times: to those on the sea-board, one-half-pint two or three times a year, to be used as medicine—not that it was always needed or used for this purpose, but too strict an inquiry into its use was wisely avoided—and for this the company demanded pay. To their servants in the interior, no liquor was sold, but they were furnished as a gratuity with one pint on leaving rendezvous, and another on arriving at winter quarters. By this management, it became impossible for them to dispose of drink to the Indians, their small allowance being always immediately consumed in a meeting or parting carouse.

The arrival of men from the interior at Fort Vancouver usually took place in the month of June, when the Columbia was high, and a stirring scene it was. The chief traders generally contrived their march through the upper country, their camps, and their rendezvous, so as to meet the Express, which annually came to Vancouver from Canada and the Red River settlements. They then descended the Columbia together and arrived in force at the Fort. This annual fleet went by the name of Brigade—a name which suggested a military spirit in the crews that their appearance failed to vindicate. Yet, though there was nothing warlike in the scene, there was much that was exciting, picturesque, and even brilliant, for these *couriers de bois*, or wood-rangers, and the *voyageurs*, or boatmen, were the most foppish of mortals when they came to rendezvous. Then, too, there was an exaltation of spirits on their safe arrival at head-quarters, after their year's toil and danger in wildernesses, among Indians and wild beasts, exposed to famine and accident, that almost deprived them of what is called *common sense*, and compelled them to the most fantastic excesses.

Their well-understood peculiarities did not make them less welcome at Vancouver. When the cry was given—"The Brigade! The Brigade!"—there was a general rush to the river's bank to witness the spectacle. In advance came the chief-trader's barge, with the company's flag at the bow, and the cross of St. George at the stern: the fleet as many abreast as the turnings of the river allowed. With strong and skillful strokes, the boatmen governed their richly laden boats, keeping them in line, and at the same time, singing in chorus a loud and not unmusical hunting or boating song. The gay ribbons and feathers with which the singers were bedecked took nothing from the picturesqueness of their appearance. The broad, full river, sparkling in the sunlight, gemmed with emerald islands, and bordered with a rich growth of flowering shrubbery. The smiling plain surrounding the Fort, and the distant mountains, where glittered the sentinel Mt. Hood, all came gracefully into the picture, and seemed to furnish a fitting background and middle distance for the bright bit of coloring given by the moving life in the scene. As with a skillful sweep, the brigade touched the bank, and the traders and men sprang on shore, and the first cheer which had welcomed their appearance was heartily repeated, while a gay clamor of questions and answers followed.

After the business immediately incident to their arrival had been dispatched, then took place the regale of pork, flour, and spirits, which was sure to end in a carouse, during which blackened eyes and broken noses were not at all uncommon. But though blood was made to flow, life was never put seriously in peril, and the belligerent parties were the best of friends when the fracas was ended.

The business of exchange being completed in three or four weeks—the rich stores of peltries consigned to their places in the warehouse, and the boats re-laden with goods for the next year's trade with the Indians in the upper country, a parting carouse took place, and with another parade of feathers, ribbons, and other finery, the brigade departed with songs and cheers as it had come, but with probably heavier hearts.

It would be a stern morality indeed which could look upon the excesses of this peculiar class as it would upon the same excesses committed by men in the enjoyment of all the comforts and pleasures of civilized life. For them, during most of the year, was only an outdoor life of toil, watchfulness, peril, and isolation. When they arrived at the rendezvous, for the brief period of their stay they were allowed perfect license because nothing else would content them. Although at headquarters they were still in the wilderness, thousands of miles from civilization, with no chance of such recreations as men in the continual enjoyment of life's sweetest pleasures would naturally seek. For them there was only one method of seeking and finding temporary oblivion of the accustomed hardship, and whatever may be the strict rendering of man's duty as an immortal being, we cannot help being somewhat lenient at times to his errors as a mortal.

After the departure of the boats, there was another arrival at the Fort of trappers from the Snake River County. Previous to 1832, such were the dangers of the fur trade in this region, that only the most experienced traders suffered to conduct a party through it, and even they were frequently attacked, and sometimes sustained serious losses of men and animals. Subsequently, however, the Hudson's Bay Company obtained such an influence over even these hostile tribes as to make it safe for a party of no more than two of their men to travel through this much-dreaded region.

There was another important arrival at Fort Vancouver, usually in midsummer. This was the Company's supply ship from London. In the possible event of a vessel being lost, one cargo was always kept on store at Vancouver, but for which wise regulation, much trouble and disaster might have resulted, especially in the early days of the establishment. Occasionally a vessel foundered at sea or was lost on the bar of the Columbia, but these losses did not interrupt the regular transaction of business. The arrival of a ship from London was the occasion of great bustle and excitement also. She brought not only goods for the posts throughout the district of the Columbia, but letters, papers, private parcels, and all that seemed of so much value to the little isolated world at the Fort.

A company conducting its business with such method and regularity as has been described, was certain of success. Yet some credit also must attach to certain individuals in its service, whose faithfulness, zeal, and ability in carrying out its designs, contributed largely to its welfare. Such a man was at the head of the Hudson's Bay Company's affairs in the large and important district west of the Rocky Mountains. The Company never had in its service a more efficient man than Gov. John McLaughlin, more commonly called Dr. McLaughlin.

To the discipline, at once severe and just, which Dr. McLaughlin maintained in his district, was due to the safety and prosperity of the company he served, and the servants of that company generally, as well as, at a later period, of the emigration which followed the hunter and trapper into the wilds of Oregon. Careful as were all the officers of the Hudson's Bay Company, they could not always avoid conflicts with the Indians, nor was their kindness and justice always sufficiently appreciated to prevent the outbreak of savage instincts. Fort Vancouver had been threatened in an early day, a vessel or two had been lost in which the Indians were suspected to have been implicated, and at long intervals a trader was murdered in the interior, or more frequently, Indian insolence put to the test both the wisdom and courage of the officers to prevent an outbreak.

When murders and robberies were committed, it was the custom at Fort Vancouver to send a strong party to demand the offenders from their tribe. Such was the well-known power and influence of the Company, and such the wholesome fear of the *King George men*, that this demand was never resisted, and if the murderer could be found he was given up to be hung according to *King George laws*. They were almost equally impelled to good conduct by the state of dependence on the company into which they had been brought. Once they had subsisted and clothed themselves from the spoils of the rivers and forest, since they had tasted of the tree of knowledge of good and evil, they could no more return to skins for raiment, nor to game alone for food. Blankets and flour, beads, guns, and ammunition had become dear to their hearts: for all these things, they must love and obey the Hudson's Bay Company. Another fine stroke of policy in the Company was to destroy the chieftain-ships in the various tribes, thus weakening them by dividing them and preventing dangerous coalitions of the leading spirits: for in savage as well as civilized life, the many are governed by the few.

It may not be uninteresting in this place to give a few anecdotes of the manner in which conflicts with the Indians were prevented, or offenses punished by the Hudson's Bay Company. In the year 1828, the ship *William and Ann* was cast away just inside the bar of the Columbia, under circumstances which seemed to direct suspicion to the Indians in that vicinity. Whether or not they had attacked the ship, not a soul was saved from the wreck to tell how she was lost. On hearing that the ship had gone to pieces, and that the Indians had appropriated a portion of her cargo, Dr. McLaughlin sent a message to the chiefs, demanding restitution of the stolen goods. Nothing was returned by the messenger except one or two worthless articles. Immediately, an armed force was sent to the scene of the robbery with a fresh demand for the goods, which the chiefs, in view of their spoils, thought proper to resist by firing upon the reclaiming party. But they were not unprepared, and a swivel was discharged to let the savages know what they might expect in the way of firearms. The argument was conclusive: the Indians fleeing into the woods. While making a search for the goods, a portion of which were found, a chief was observed skulking near, and cocking his gun, on which motion one of the men fired, and he fell. This prompt action, the justice of which the Indians well understood, and the intimidating power of the swivel put an end to the incipient war. Care was then taken to impress upon their minds that they must not expect to profit by the disasters of vessels, nor be tempted to murder white men for the sake of plunder. The *William and Ann* was supposed to have got aground, when the savages, seeing her situation, boarded her and murdered the crew for the cargo which they knew her to contain. Yet, as there were no positive proofs, only such measures were taken as would deter them from a similar attempt in the future. That the lesson was not lost, was proven two years later, when the *Isabella*, from London, struck on the bar, her crew deserting her. In this instance, no attempt was made to meddle with the vessel's cargo, and as the crew made their way to Vancouver, the goods were nearly all saved.

In a former voyage of the *William and Ann* to the Columbia River, she had been sent on an exploring expedition to the Gulf of Georgia to discover the mouth of Frazier's River, having on board a crew of forty men. Whenever the ship came to anchor, two sentries were kept constantly on deck to guard against any surprise or misconduct on the part of the Indians, so adroit, however, were they in the light-fingered art, that every one of the eight cannon with which the ship was armed was robbed of its ammunition, as was discovered on leaving the river! Such incidents as these served to impress the minds of the Company's officers and servants with the necessity of vigilance in their dealings with the savages.

Not all their vigilance could at all times avail to prevent mischief. When Sir George Simpson, Governor of the Hudson's Bay Company, was on a visit to Vancouver in 1829, he was made aware of this truism. The governor was on his return to Canada by way of the Red River Settlement and had reached the Dalles of the Columbia with his party. In making the portage at this place, all the party except Dr. Tod gave their guns into the charge of two men to prevent their being stolen by the Indians, who crowded about, and whose well-known bad character made great care needful. All went well with no attempt to seize either guns or other property being made until at the end of the portage, the boats had been reloaded. As the party were about to re-embark, a simultaneous rush was made by the Indians who had dogged their steps, to get possession of the boats. Dr. Tod raised his gun immediately, aiming at the head chief, who, not liking the prospect of so speedy dissolution, ordered his followers to desist, and the party suffered to escape. It was soon after discovered that every gun belonging to the party in the boat had been wet, except the one carried by Dr. Tod, and to the fact that the doctor did carry his gun, all the others owed their lives.

The great desire of the Indians for guns and ammunition led to many stratagems which were dangerous to the possessors of the coveted articles. Much more dangerous would it have been to have allowed them a free supply of these things, nor could an Indian purchase from the Company more than a stated supply, which was to be used, not for the purposes of war, but to keep himself in the game.

Dr. McLaughlin was himself once quite near falling into a trap of the Indians, so cunningly laid as to puzzle even him. This was a report brought to him by a deputation of Columbia River Indians, stating the startling fact that the fort at Nisqually had been attacked, and every inmate slaughtered. To this horrible story, told with every appearance of truth, the doctor listened with incredulity mingled with apprehension. The Indians were closely questioned and cross-questioned but did not conflict in their testimony. The matter assumed a very painful aspect. Not to be deceived, the doctor had the unwelcome messengers committed to custody while he could bring other witnesses from their tribe. But they were prepared for this, and the whole tribe was as positive as those who brought the tale. Confounded by this cloud of witnesses, Dr. McLaughlin had almost determined upon sending an armed force to Nisqually to inquire into the matter, and if necessary, punish the Indians, when a detachment of men arrived from that post, and the plot was exposed! The design of the Indians had been simply to cause a division of the force at Vancouver, after which they believed they might succeed in capturing and plundering the fort. Had they truly been successful in this undertaking, every other trading post in the country would have been destroyed. But so long as the headquarters of the Company remained secure and powerful, the other stations were comparatively safe.

An incident which has been several times related, occurred at Fort Walla-Walla, and shows how narrow escapes the interior traders sometimes made. The hero of this anecdote was Mr. McKinlay, one of the most estimable of the Hudson's Bay Company's officers, in charge of the fort just named. An Indian was one day lounging about the fort and seeing some timbers lying in a heap that had been squared for pack saddles, helped himself to one and commenced cutting it down into a whip handle for his own use. To this procedure, Mr. McKinlay's clerk demurred, first telling the Indian its use, and then ordering him to resign the piece of timber. The Indian insolently replied that the timber was his, and he should take it. At this, the clerk, with more temper than prudence, struck the offender, knocking him over, soon after which the savage left the fort with sullen looks, boding vengeance. The next day, Mr. McKinlay, not being informed of what had taken place, was in a room of the fort with his clerk when a considerable party of Indians began dropping quietly in until there were fifteen or twenty of them inside the building. The first intimation of anything wrong McKinlay received was when he observed the clerk pointed out in a particular manner by one of the party. He instantly comprehended the purpose of his visitors, and with that quickness of thought which is habitual to the student of savage nature, he rushed into the storeroom and returned with a powder keg, flint, and steel. By this time the unlucky clerk was struggling for his life with his vindictive foes. Putting down the powder in their midst and knocking out the head of the keg with a blow, McKinlay stood over it ready to strike fire with his flint and steel. The savages paused, aghast. They knew the nature of the *perilous stuff*, and also understood the trader's purpose. "Come," said he with a clear, determined voice, "you are twenty braves against us two: now touch him if you dare, and see who dies first." In a moment, the fort was cleared, and McKinlay was left to inquire the cause of what had so nearly been a tragedy. It is hardly a subject of doubt whether or not his clerk got a scolding. Soon after, such was the powerful influence exerted by these gentlemen, the chief of the tribe flogged the pilfering Indian for the offense, and McKinlay became a great brave, a *big heart* for his courage.

It was indeed necessary to have courage, patience, and prudence in dealing with the Indians. These the Hudson's Bay officers generally possessed. Perhaps the most irascible of them all in the Columbia District, was their chief, Dr. McLaughlin, but such was his goodness and justice that even the savages recognized it, and he was *hyas tyee*, or great chief, in all respects to them. Being on one occasion very much annoyed by the pertinacity of an Indian who was continually demanding pay for some stones with which the doctor was having a vessel ballasted, he seized one of some size, and thrusting it in the Indian's mouth, cried out in a furious manner, "Pay, pay! If the stones are yours, take them and eat them, you rascal! Pay, pay! The devil! The devil!" Upon which explosion of wrath, the native owner of the soil thought it prudent to withdraw his immediate claims.

There was more, however, in the doctor's action than mere indulgence of wrath. He understood perfectly that the savage values only what he can eat and wear, and that as he could not put the stones to either of these uses, his demand for pay was an impudent one.

Enough has been said to give the reader an insight into Indian character, to prepare his mind for events which are to follow, to convey an idea of the influence of the Hudson's Bay Company, and to show on what it was founded. The American fur companies will now be sketched, and their mode of dealing with the Indians contrasted with that of the British Company. The comparison will not be favorable, but should any unfairness be suspected, a reference to Mr. Irving's *Bonneville*, will show that the worthy captain was forced to witness against his own countrymen in his narrative of his hunting and trading adventures in the Rocky Mountains.

The dissolution of the Pacific Fur Company, the refusal of the United States Government to protect Mr. Astor in a second attempt to carry on a commerce with the Indians west of the Rocky Mountains, and the occupation of that country by British traders, had the effect to deter individual enterprise from again attempting to establish commerce on the Pacific coast. The people waited for the government to take some steps toward the encouragement of transcontinental trade, the government beholding the lion—British—in the way, waiting for the expiration of the convention of 1818 in the Micawber-like hope that something would *turn up* to settle the question of territorial sovereignty. The war of 1812 had begun on the part of Great Britain, to secure the great western territories to herself for the profits of the fur trade, almost solely. Failing in this, she had been compelled, by the treaty of Ghent, to restore to the United States all the places and forts captured during that war. Yet the forts and trading posts in the west remained practically in the possession of Great Britain, for her traders and fur companies still roamed the country, excluding American trade, and inciting —so the frontiersmen believed—the Indians to acts of blood and horror.

Congress, being importuned by the people of the West, finally, in 1815, passed an act expelling British traders from American territory east of the Rocky Mountains. Following the passage of this act, the hunters and trappers of the old North American Company, at the head of which Mr. Astor still remained, began to range the country about the headwaters of the Mississippi and the upper Missouri. Also, a few American traders had ventured into the northern provinces of Mexico, previous to the overthrow of the Spanish Government, and after that event, a thriving trade grew up between St. Louis and Santa Fe.

At length, in 1823, Mr. W. H. Ashley[2], of St. Louis, a

2. Ashley made his trips beyond the Rockies in 1824 and 1825, not 1823 and 1824. Credit for the explorations mentioned belongs to his men, especially Jedediah Smith, Tom Fitzpatrick and Jim Bridger, not to Ashley personally. He built no fort in the Salt Lake region. He sold his business in 1826, not 1827, to Smith, Jackson & Sublette, who did not use the name Rocky Mountain Fur Company.

merchant for a long time engaged in the fur trade on the Missouri and its tributaries, determined to push a trading party up to or beyond the Rocky Mountains. Following up the Platte River, Mr. Ashley proceeded at the head of a large party with horses and merchandise, as far as the northern branch of the Platte, called the Sweetwater. This he explored to its source, situated in that remarkable depression in the Rocky Mountains, known as the South Pass—the same which Fremont *discovered* twenty years later, during which twenty years it was annually traveled by trading parties, and just prior to Fremont's discovery, by missionaries and emigrants destined to Oregon. To Mr. Ashley also belongs the credit of having first explored the headwaters of the Colorado, called the Green River, afterward a favorite rendezvous of the American fur companies. The country about the South Pass proved to be an entirely new hunting ground, and very rich in furs, as here many rivers take their rise, whose headwaters furnished abundant beaver. Here, Mr. Ashley spent the summer, returning to St. Louis in the fall with a valuable collection of skins.

In 1824, Mr. Ashley repeated the expedition, extending it this time beyond Green River as far as the Great Salt Lake, near which, to the south, he discovered another smaller lake, which he named Lake Ashley, after himself. On the shores of this lake, he built a fort for trading with the Indians, and leaving in it about one hundred men, returned to St. Louis the second time with a large amount of furs. During the time the fort was occupied by Mr. Ashley's men, a period of three years, more than one hundred and eighty thousand dollars worth of furs were collected and sent to St. Louis. In 1827, the fort, and all of Mr. Ashley's interest in the business was sold to the Rocky Mountain

Fur Company, at the head of which were Jedediah Smith, William Sublette, and David Jackson, Sublette being the leading spirit in the Company.

The custom of these enterprising traders, who had been in the mountains since 1824, was to divide their force, each taking his command to a good hunting ground, and returning at stated times to rendezvous, generally appointed on the headwaters of Green River. Frequently, the other fur companies—for there were other companies formed on the heels of Ashley's enterprise —learning of the place appointed for the yearly rendezvous, brought their goods to the same resort, when an intense rivalry was exhibited by the several traders as to which company should soonest dispose of its goods, getting, of course, the largest amount of furs from the trappers and Indians. So great was the competition in the years between 1826 and 1829, when there were about six hundred American trappers in and about the Rocky Mountains, besides those of the Hudson's Bay Company, that it was death for a man of one company to dispose of his furs to a rival association. Even a *free trapper*—that is, one not indentured, but hunting upon certain terms of agreement concerning the price of his furs and the cost of his outfit, only, dared not sell to any other company than the one he had agreed with.

Jedediah Smith[3], of the Rocky Mountain Fur Company, during their first year in the mountains, took a party of five trap-

3. Smith made two trips to California, in 1826 and 1827-28, not starting in 1827. His route on the first trip was from the vicinity of Salt Lake southwest to the vicinity of Los Angeles, not from Santa Fe to San Francisco Bay. He traveled north along the coast to the Columbia on his second trip, not his first, and had his conflict with the Umpquas after his conflict with the Mojave Indians on the Colorado River, not before. And neither of these conflicts was after the time Meek went to the mountains, as implied.

into Oregon, being the first American, trader or other, to cross into that country since the breaking up of Mr. Astor's establishment. He was trapped on the headwaters of the Snake River until autumn, when he fell in with a party of Hudson's Bay trappers, and going with them to their post in the Flathead country, wintered there.

Again, in 1826, Smith, Sublette, and Jackson, brought out a large number of men to trap in the Snake River country, and entered into direct competition with the Hudson's Bay Company, whom they opposed with hardly a degree more of zeal than they competed with rival American traders: this one extra degree being inspired by a *spirit of '76* toward anything British.

After the Rocky Mountain Fur Company had extended its business by the purchase of Mr. Ashley's interest, the partners determined to push their enterprise to the Pacific coast, regardless of the opposition they were likely to encounter from the Hudson's Bay traders. Accordingly, in the spring of 1827, the Company was divided up into three parts, to be led separately by different routes into the Indian Territory, nearer the ocean.

Smith's route was from the Platte River, southward to Santa Fe, thence to the bay of San Francisco, and thence along the coast to the Columbia River. His party was successful and had arrived in the autumn of the following year at the Umpqua River, about two hundred miles south of the Columbia, in safety. Here, one of those sudden reverses to which the *mountain man* is liable at any moment, overtook him. His party at this time consisted of thirteen men with their horses and a collection of furs valued at twenty thousand dollars. Arrived at the Umpqua, they encamped for the night on its southern bank, unaware that the natives in this vicinity—the Shastas—were fiercer and more treacherous than the indolent tribes of California, for whom, probably, they had a great contempt. All went well until the following morning, the Indians hanging about the camp, but apparently friendly. Smith had just breakfasted and was occupied in looking for a fording-place for the animals, being on a raft, and having with him a little Englishman and one Indian. When they were in the middle of the river, the Indian snatched Smith's gun and jumped into the water. At the same instant, a yell from the camp, which was in sight, proclaimed that it was attacked. Quick as thought, Smith snatched the Englishman's gun and shot dead the Indian in the river.

To return to the camp was certain death. Already, several of his men had fallen, and overpowered by numbers, he could not hope that any would escape, and nothing was left him but flight. He succeeded in getting to the opposite shore with his raft before he could be intercepted, and fled with his companion, on foot and with only one gun, and no provisions, to the mountains that border the river. With great good fortune they were enabled to pass through the remaining two hundred miles of their journey without accident, though not without suffering, and reach Fort Vancouver in a destitute condition, where they were kindly cared for.

Of the men left in camp, only two escaped. One man named Black defended himself until he saw an opportunity for flight, when he escaped to the cover of the woods, and finally to a friendly tribe farther north, near the coast, who piloted him to Vancouver. The remaining man was one Turner, of a very powerful frame, who was doing camp duty as a cook on this eventful morning. When the Indians rushed upon him, he defended himself with a huge firebrand, or half-burned poplar stick, with which he laid about him like Sampson, killing four red skins before he saw a chance of escape. Singularly, for one, in his extremity, he did escape and also arrived in Vancouver that winter.

Dr. McLaughlin received the unlucky trader and his three surviving men with every mark and expression of kindness and entertained them through the winter. Not only this, but he dispatched a strong, armed party to the scene of the disaster to punish the Indians and recover the stolen goods, all of which was done at his own expense, both as an act of friendship toward his American rivals, and as necessary to the discipline which they everywhere maintained among the Indians. Should this offense go unpunished, the next attack might be upon one of his own parties going annually down into California. Sir George Simpson, the governor of the Hudson's Bay Company, chanced to be spending the winter at Vancouver. He offered to send Smith to London the following summer, in the Company's vessel, where he might dispose of his furs to advantage, but Smith declined this offer and finally sold his furs to Dr. McLaughlin and returned in the spring to the Rocky Mountains.

On Sublette's return from St. Louis, in the summer of 1829, with men and merchandise for the year's trade, he became uneasy on account of Smith's protracted absence. According to a previous plan, he took a large party into the Snake River country to hunt. Among the recruits from St. Louis was Joseph L. Meek, the subject of the narrative following this chapter. Sublette, not meeting with Smith's party on its way from the Columbia, as he still hoped, at length detailed a party to look for him on the headwaters of the Snake. Meek was one of the men sent to look for the missing partner, whom he discovered at length in Pierre's

Hole, a deep valley in the mountains, from which issues the Snake River in many living streams. Smith returned with the men to camp, where the tale of his disasters was received after the manner of mountain men, simply declaring with a momentarily sobered countenance that their comrade has not been *in luck*, with which brief and equivocal expression of sympathy the subject is dismissed. To dwell on the dangers incident to their calling would be to half disarm themselves of their necessary courage, and it is only when they are gathered about the fire in their winter camp, that they indulge in tales of wild adventure and *hairbreadth 'scapes*, or make sorrowful reference to a comrade lost.

Influenced by the hospitable treatment which Smith had received at the hands of the Hudson's Bay Company, the partners now determined to withdraw from competition with them in the Snake country, and to trap upon the waters of the Colorado, in the neighborhood of their fort. But *luck*, the mountain man's Providence, seemed to have deserted Smith. In crossing the Colorado River with a considerable collection of skins, he was again attacked by Indians and only escaped by losing all his property. He then went to St. Louis for a supply of merchandise and fitted out a trading party for Santa Fe, but on his way to that place, he was killed in an encounter with the savages.[4]

Turner, the man who so valiantly wielded the firebrand on the Umpqua River, several years later met with a similar adven-

4. Smith, Jackson & Sublette sold out to Bridger, Fitzpatrick, Milton Sublette, Fraeb, and Gervais in 1830, before Smith's death. These new partners were the first to use the name Rocky Mountain Fur Company.

on the Rogue River, in Southern Oregon, and was the means of saving the lives of his party by his courage, strength, and alertness. He finally, when trapping had become unprofitable, retired upon a farm in the Willamette Valley, as did many other mountain men who survived the dangers of their perilous trade.

After the death of Smith, the Rocky Mountain Fur Company continued its operations under the command of Bridger, Fitzpatrick, and Milton Sublette, brother of William. In the spring of 1830, they received about two hundred recruits, and with little variation, kept up their number of three or four hundred men for a period of eight or ten years longer, or until the beaver were hunted out of every nook and corner of the Rocky Mountains.

Previous to 1835, there were in and about the Rocky Mountains, beside the *American* and *Rocky Mountain* companies, the St. Louis Company, and eight or ten *lone traders*. Among these latter were William Sublette, Robert Campbell, J. O. Pattie, Mr. Pilcher, Col. Charles Bent, St. Vrain, William Bent, Mr. Gant, and Mr. Blackwell. All these companies and traders more or less frequently penetrated into the countries of New Mexico, Old Mexico, Sonora, and California, returning sometimes through the mountain regions of the latter State, by the Humboldt River to the headwaters of the Colorado. Seldom, in all their journeys, did they intrude on that portion of the Indian Territory lying within three hundred miles of Fort Vancouver, or which forms the area of the present State of Oregon.

Up to 1832, the fur trade in the West had been chiefly conducted by merchants from the frontier cities, especially by those of St. Louis. The old *North American* was the only exception. But in the spring of this year, Captain Bonneville, a United States officer on furlough, led a company of a hundred men, with a train of wagons, horses and mules, with merchandise, into the trapping grounds of the Rocky Mountains. His wagons were the first that had ever crossed the summit of these mountains, though William Sublette had, two or three years previous, brought wagons as far as the valley of the Wind River on the east side of the range. Captain Bonneville remained nearly three years in the hunting and trapping grounds, taking parties of men into the Colorado, Humboldt, and Sacramento valleys, but he realized no profits from his expedition, being opposed and competed with by both British and American traders of larger experience.

But Captain Bonneville's venture was a fortunate one compared with that of Mr. Nathaniel Wyeth of Massachusetts, who also crossed the continent in 1832, with the view of establishing a trade on the Columbia River. Mr. Wyeth brought with him a small party of men, all inexperienced in frontier or mountain life, and destined for a salmon fishery on the Columbia. He had reached Independence, Missouri, the last station before plunging into the wilderness, and found himself somewhat at a loss how to proceed, until, at this juncture, he was overtaken by the party of William Sublette, from St. Louis to the Rocky Mountains, with whom he traveled in company to the rendezvous at Pierre's Hole.

When Wyeth arrived at the Columbia River, after tarrying until he had acquired some mountain experiences, he found that his vessel, which was loaded with merchandise for the Columbia River trade, had not arrived. He remained at Vancouver through the winter, the guest of the Hudson's Bay Company, and either having learned or surmised that his vessel was wrecked, returned to the United States in the following year. Not discouraged, however, he made another venture in 1834, dispatching the ship *May Dacre*, Captain Lambert, for the Columbia River, with another cargo of Indian goods, traveling himself overland with a party of two hundred men, and a considerable quantity of merchandise which he expected to sell to the Rocky Mountain Fur Company. In this expectation, he was defeated by William Sublette, who had also brought out a large assortment of goods for the Indian trade, and had sold out, supplying the market, before Mr. Wyeth arrived.

Wyeth then built a post, named Fort Hall, on Snake River, at the junction of the Portneuf, where he stored his goods, and having detached most of his men in trapping parties, proceeded to the Columbia River to meet the *May Dacre*. He reached the Columbia about the same time with his vessel and proceeded at once to erect a salmon fishery. To forward this purpose, he built a post, called Fort William, on the lower end of Wappatoo—now known as Sauvie's—Island, near where the Lower Wallamet falls into the Columbia. But for various reasons, he found the business on which he had entered unprofitable. He had much trouble with the Indians, his men were killed or drowned, so that by the time he had half a cargo of fish, he was ready to abandon the effort to establish a commerce with the Oregon Indians, and was satisfied that no enterprise less stupendous and powerful than that of the Hudson's Bay Company could be long sustained in that country.

Much complaint was subsequently made by Americans, chiefly missionaries, of the conduct of that company in not allowing Mr. Wyeth to purchase beaver skins of the Indians, hut Mr. Wyeth himself made no such complaint. Personally, he was treated with unvarying kindness, courtesy, and hospitality. As a trader, they would not permit him to undersell them. In truth, they no doubt wished him away, because competition would soon ruin the business of either, and they liked not to have the Indians taught to expect more than their furs were worth, nor to have the Indians' confidence in themselves destroyed or tampered with.

The Hudson's Bay Company were hardly so unfriendly to him as the American companies, since to the former he was enabled to sell his goods and fort on the Snake River, before he returned to the United States, which he did in 1835.

The sale of Fort Hall to the Hudson's Bay Company was a finishing blow at the American fur trade in the Rocky Mountains, which after two or three years of constantly declining profits, was entirely abandoned.

Something of the dangers incident to the life of the hunter and trapper may be gathered from the following statements, made by various parties who have been engaged in it. In 1808, a Missouri Company engaged in fur hunting on the three forks of the river Missouri, were attacked by Blackfeet, losing twenty-seven men, and being compelled to abandon the country. In 1823, Mr. Ashley was attacked on the same river by the Arickaras and had twenty-six men killed. About the same time the Missouri company lost seven men and fifteen thousand dollars' worth of merchandise on the Yellowstone River. A few years previous, Major Henry lost, on the Missouri River, six men and fifty horses. In the sketch given of Smith's trading adventures is shown how uncertain were life and property at a later period. Of the two hundred men whom Wyeth led into the Indian country, only about forty were alive at the end of three years. There was, indeed, a constant state of warfare between the

Indians and the whites, wherever the American companies hunted, in which great numbers of both lost their lives. Add to this cause of decimation the perils from wild beasts, famine, cold, and all manner of accidents, and the trapper's chance of life was about one in three.[5]

Of the causes which have produced the enmity of the Indians, there are about as many. It was found to be the case almost universally, that on the first visit of the whites, the natives were friendly after their natural fears had been allayed. But by degrees, their cupidity was excited to possess themselves of the much-coveted dress, arms, and goods of their visitors. As they had little or nothing to offer in exchange, which the white man considered an equivalent, they took the only method remaining of gratifying their desire of possession and *stole* the coveted articles which they could not purchase. When they learned that the white men punished theft, they murdered to prevent the punishment. Often, also, they had wrongs of their own to avenge. White men did not always regard their property-rights. They were guilty of infamous conduct toward Indian women. What one party of whites told them was true, another plainly contradicted, leaving the lie between them. They were overbearing toward the Indians on their own soil, exciting to irrepressible hostility the natural jealousy of the inferior toward the superior race, where both are free, which characterizes all people. In short, the Indians were not without their grievances, and from barbarous ignorance and wrong on one side, and intelligent

5. Mrs. Victor exaggerates the risk to life the mountain men ran. The statistical studies in Volume I of the Hafen series show that her estimate that the chance of staying alive was one in three is wildly inaccurate.

wrong doing on the other, together with the misunderstandings likely to arise between two entirely distinct races, grew constantly a thousand abuses, which resulted in a deadly enmity between the two.

For several reasons this evil existed to a greater degree among the American traders and trappers than among the British. The American trapper was not, like the Hudson's Bay employees, bred to the business. Oftener than any other way, he was some wild youth who, after an *escapade* in the society of his native place, sought safety from reproach or punishment in the wilderness. Or he was some disappointed man who, with feelings embittered toward his fellows, preferred the seclusion of the forest and mountain. Many were of a class disreputable everywhere, who gladly embraced a life not subject to social laws. A few were brave, independent, and hardy spirits, who delighted in the hardships and wild adventures their calling made necessary. All these men, the best with the worst, were subject to no will but their own, and all experience goes to prove that a life of perfect liberty is apt to degenerate into a life of license. Even their own lives, and those of their companions, when it depended upon their own prudence, were but lightly considered. The constant presence of danger made them reckless. It is easy to conceive how, under these circumstances, the natives and the foreigners grew to hate each other, in the Indian country, especially after the Americans came to the determination to *shoot an Indian at sight*, unless he belonged to some tribe with whom they had intermarried, after the manner of the trappers.[6]

6. The American trappers were not such a majority of scoundrels as Mrs. Victor suggests.

On the other hand, the employees of the Hudson's Bay Company were many of them half-breeds or full-blooded Indians of the Iroquois nation, toward whom nearly all the tribes were kindly disposed. Even the Frenchmen who trapped for this company were well-liked by the Indians on account of their suavity of manner, and the ease with which they adapted themselves to savage life. Besides, most of them had native wives and half-breed children, and were regarded as relatives. They were trained to the life of a trapper, were subject to the will of the Company, and were generally just and equitable in their dealings with the Indians, according to that company's will, and the dictates of prudence. Here was a wide difference.

Notwithstanding this, there were many dangers to be encountered. The hostility of some of the tribes could never be overcome, nor has it ever abated. Such were the Crows, the Blackfeet, the Cheyennes, the Apaches, and the Comanches. Only a superior force could compel the friendly offices of these tribes for any white man, and then their treachery was as dangerous as their open hostility.

It happened, therefore, that although the Hudson's Bay Company lost comparatively few men by the hands of the Indians, they sometimes found them implacable foes in common with the American trappers, and frequently one party was very glad of the others' assistance. Altogether, as has before been stated, the loss of life was immense in proportion to the number employed.

Very few of those who had spent years in the Rocky Mountains ever returned to the United States. With their Indian wives and half-breed children, they scattered themselves throughout

Oregon, until when, a number of years after the abandonment of the fur trade, Congress donated large tracts of land to actual settlers, they laid claim, each to his selected portion, and became active citizens of their adopted state.

# The River of the West, The Mountain Years

# 1

As has been stated in the Introduction, Joseph L. Meek was a native of Washington Co., Va. Born in the early part of the present century, and brought up on a plantation where the utmost liberty was accorded to the *young massa*, preferring outdoor sports with the youthful bondsmen of his father, to study with the bald-headed schoolmaster who furnished him the alphabet on a paddle—possessing an exhaustless fund of waggish humor, united to a spirit of adventure and remarkable personal strength, he unwittingly furnished in himself the very material of which the heroes of the wilderness were made. Virginia, *the mother of Presidents*, has furnished many such men, who, in the early days of the now populous Western States, became the hardy frontiersmen, or the fearless Indian fighters who were the bone and sinew of the land.

When young Joe was about eighteen years of age, he wearied of the monotony of plantation life, and jumping into the wagon of a neighbor who was going to Louisville, Ky., started out in life for himself. He *reckoned they did not grieve for him at home*, at which conclusion others besides Joe naturally arrive on hearing of his heedless disposition and utter contempt for the ordinary and useful employments to which other men apply themselves. This truly Virginian and chivalric contempt for *honest labor* has continued to distinguish him throughout his eventful career, even while performing the most arduous duties of the life he had chosen.

Joe probably believed that should his father grieve for him, his stepmother would be able to console him. This stepmother, though a pious and good woman, not being one of the lad's favorites, as might easily be conjectured. It was such thoughts as these that kept up his resolution to seek the far west. In the autumn of 1828, he arrived in St. Louis, and the following spring, he fell in with Mr. WM. Sublette, of the Rocky Mountain Fur Company, who was making his annual visit to that frontier town to purchase merchandise for the Indian country and pick up recruits for the fur-hunting service. To this experienced leader he offered himself.

"How old are you?" asked Sublette.

"A little past eighteen."

"And you want to go to the Rocky Mountains?"

"Yes."

"You don't know what you are talking about, boy. You'll be killed before you get halfway there."

"If I do, I reckon I can die!" said Joe, with a flash of his full dark eyes, and throwing back his shoulders to show their breadth.

"Come," exclaimed the trader, eyeing the youthful candidate with admiration and perhaps a touch of pity also. "That is the game spirit. I think you'll do, after all. Only be prudent and keep your wits about you."

"Where else should they be?" Joe laughed as he marched off, feeling an inch or two taller than before.

Then commenced the business of preparing for the journey —making acquaintance with the other recruits—enjoying the novelty of owning an outfit, being initiated into the mysteries of camp duty by the few old hunters who were to accompany the expedition and learning something of their swagger and disregard of civilized observances.

On the 17th of March 1829, the company, numbering about sixty men, left St. Louis and proceeded on horses and mules, with packhorses for the goods, up through the state of Missouri. Camp life commenced at the start, and this being the season of the year when the weather is most disagreeable, its romance rapidly melted away with the snow and sleet, which varied the sharp spring wind and the frequent cold rains. The recruits went through all the little mishaps incident to the business and to their inexperience, such as involuntary somersaults over the heads of their mules, bloody noses, bruises, dusty faces, bad colds, accidents in fording streams, yet withal no very serious hurts or hindrances. Rough weather and severe exercise gave them wolfish appetites, which sweetened the coarse camp-fare and amateur cooking.

Getting up at four o'clock of a March morning to kindle fires and attend to the animals was not the most delectable duty that our labor-despising young recruit could have chosen. But, if he repented of the venture he had made, nobody was the wiser. Sleeping off stormy nights in corn cribs or under sheds,sheds could not be by any stretch of the imagination converted into a highly romantic or heroic mode of lodging oneself. The squalid manner of living of the few inhabitants of Missouri at this period gave a forlorn aspect to the country which is lacking in the wilderness itself—a thought which sometimes occurred to Joe like a hope for the future. Mountain-fare, he began to think, must be better than the boiled corn and pork of the Missourians. Antelope and buffalo meat were more suitable viands for a hunter than coon and opossum. Thus those very duties which seemed undignified, and those hardships without danger or glory, which marked the beginning of his career made him ambitious of a more free and hazardous life on the plains and in the mountains.

Among the recruits was a young man not far from Joe's own age, named Robert Newell, from Ohio. One morning, when the company was encamped near Boonville, the two young men were out looking for their mules, when they encountered an elderly woman returning from the milking yard with a gourd of milk. Newell made some remark on the style of vessel she carried, when she broke out in a sharp voice.

"Young chap, I'll bet you run off from your mother! Who'll mend them holes in the elbow of your coat? You're a purty looking chap to go to the mountains, among them Injuns! They'll *kill* you. You'd better go back home!"

Considering that these frontier people knew what Indian fighting was, this was no doubt sound and disinterested advice, notwithstanding it was given somewhat sharply. And so the young men felt it to be, but it was not in the nature of either of them to turn back from a course because there was danger in it. The thought of home and somebody to mend their coats was, however, for the time strongly presented. But the company moved on, with undiminished numbers, stared at by the few inhabitants and having their own little adventures, until they came to Independence, the last station before committing themselves to the wilderness.

At this place, which contained a dwelling-house, cotton-gin, and grocery, the camp tarried for a few days to adjust the packs and prepare for a final start across the plains. On Sunday, the settlers got together for a shooting match, which some of the travelers joined, without winning many laurels. Coon-skins, deer-skins, and bees-wax changed hands freely among the settlers, whose skill with the rifle was greater than their hoard of silver dollars. This was the last vestige of civilization which the company could hope to behold for years, and rude as it was, yet won from them many a parting look as they finally took their way across the plains toward the Arkansas River.

Often, on this part of the march, a dead silence fell upon the party, which remained unbroken for miles of the way. Many, no doubt, were regretting homes by them abandoned, or wondering dreamily how many and whom of that company would ever see the Missouri country again. Many indeed went the way the woman of the gourd had prophesied, but not the hero of this story, nor his comrade Newell.

The route of Captain Sublette led across the country from near the mouth of the Kansas River to the River Arkansas[1], thence to the south fork of the Platte, thence on to the north fork of that River, to where Ft. Laramie now stands, thence up the

1. Sublette's route to the mountains was a typical Oregon Trail route. The reference to the Arkansas River must be an error. Likely the Kansas River was meant.

north fork to the Sweetwater, and thence across in a still northwesterly direction to the head of Wind River.

The manner of camp travel is now so well known through the writings of Irving, and still more from the great numbers which have crossed the plains since *Astoria* and *Bonneville* were written, that it would be superfluous here to enter upon a particular description of a train on that journey. A strict half-military discipline had to be maintained, regular duties assigned to each person, precautions taken against the loss of animals either by straying or Indian stampeding, etc. Some of the men were appointed as camp-keepers, who had all these things to look after, besides standing guard. A few were selected as hunters, and these were free to come and go, as their calling required. None but the most experienced were chosen for hunters, on a march and therefore, our recruit could not aspire to that dignity yet.

The first adventure the company met with worthy of mention after leaving Independence, was in crossing the country between the Arkansas and the Platte. Here, the camp was surprised one morning by a band of Indians a thousand strong, that came sweeping down upon them in such warlike style that even Captain Sublette was fain to believe it his last battle. Upon the open prairie, there is no such thing as flight, nor any cover under which to conceal a party even for a few moments. It is always fight or die, if the assailants are in the humor for war.

Happily, on this occasion, the band proved to be more peaceably disposed than their appearance indicated, being the warriors of several tribes—the Sioux, Arapahos, Kiowas, and Cheyennes, who had been holding a council to consider probably what mischief they could do to some other tribes. The spectacle they presented as they came at full speed on horseback, armed, painted, brandishing their weapons, and yelling in first-rate Indian style, was one which might well strike with a palsy the stoutest heart and arm. What were a band of sixty men against a thousand armed warriors in full fighting trim, with spears, shields, bows, battle-axes, and not a few guns?

But it is the rule of the mountain men to *fight*—and that there is a chance for life until the breath is out of the body—therefore Captain Sublette had his little force drawn up in line of battle. On came the savages, whooping and swinging their weapons above their heads. Sublette turned to his men. "When you hear my shot, then fire." Still, they came on, until within about fifty paces of the line of waiting men. Sublette turned his head and saw his command with their guns all up to their faces, ready to fire, then raised his own gun. Just at this moment, the principal chief sprang off his horse and laid his weapon on the ground, making signs of peace. Then followed a talk, and after the giving of a considerable present, Sublette was allowed to depart. This he did with all dispatch, the company putting as much distance as possible between themselves and their visitors before making their next camp. Considering the warlike character of these tribes and their superior numbers, it was as narrow an escape on the part of the company as it was an exceptional freak of generosity on the part of the savages to allow it. But Indians have all a great respect for a man who shows no fear, and it was most probably the warlike movement of Captain Sublette and his party which inspired a willingness on the part of the chief to accept a present, when he had the power to have taken the whole train. Besides, according to Indian logic, the present cost him nothing, and it might cost him many warriors to capture the train. Had there been the least wavering on Sublette's part, or fear in the countenances of his men, the end of the affair would have been different. This adventure was a grand initiation of the raw recruits, giving them both an insight into savage modes of attack, and an opportunity to test their own nerve.

The company proceeded without accident, and arrived, about the first of July, at the rendezvous, which was appointed for this year on the Popo Agie, one of the streams which form the headwaters of Bighorn River.

Now, indeed, young Joe had an opportunity of seeing something of the life upon which he had entered. As customary, when the traveling partner arrived at rendezvous with the year's merchandise, there was a meeting of all the partners, if they were within reach of the appointed place. On this occasion Smith was absent on his tour through California and Western Oregon, as has been related in the prefatory chapter. Jackson, the resident partner and commander for the previous year, was not yet in, and Sublette had just arrived with the goods from St. Louis.

All the different hunting and trapping parties and Indian allies were gathered together so that the camp contained several hundred men with their riding and packhorses. Nor were Indian women and children wanting to give variety and an appearance of domesticity to the scene.

The Summer rendezvous was always chosen in some valley where there was grass for the animals, and game for the camp. The plains along the Popo Agie, besides furnishing these necessary bounties, were bordered by picturesque mountain ranges, whose naked bluffs of red sandstone glowed in the morning and evening sun with a mellowness of coloring charming to the eye of the Virginia recruit. The waving grass of the plain, variegated with wildflowers, the clear summer heavens flecked with white clouds that threw soft shadows in passing, the grazing animals scattered about the meadows, the lodges of the *Booshways*[2], around which clustered the camp in motley garb and brilliant coloring, gay laughter, and the murmur of soft Indian voices, all made up a most spirited and enchanting picture, in which the eye of an artist could not fail to delight.

But as the goods were opened, the scene grew livelier. All were eager to purchase, most of the trappers to the full amount of their year's wages, and some of them, generally free trappers, went into debt to the company to a very considerable amount, after spending the value of a year's labor, privation, and danger, at the rate of several hundred dollars in a single day.

The difference between a hired and a free trapper was greatly in favor of the latter. The hired trapper was regularly indentured and bound not only to hunt and trap for his employers, but also to perform any duty required of him in camp. The Booshway, or the trader, or the partisan—leader of the detachment—had him under his command, to make him take charge of, load and unload the horses, stand guard, cook, hunt fuel, or, in short, do any and every duty. In return for this toilsome service, he received an outfit of traps, arms and ammunition, horses, and whatever his service required. Besides his outfit, he received no more than three or four hundred dollars a year as wages.

There was also a class of free trappers, who were furnished

---

2. Leaders or chiefs—corrupted from the French of Bourgeois and borrowed from the Canadians.

with their outfit by the company they trapped for, and who were obliged to agree to a certain stipulated price for their furs before the hunt commenced. But the genuine free trapper regarded himself as greatly the superior of either of the foregoing classes. He had his own horses and accouterments, arms and ammunition. He took what route he thought fit, hunted and trapped when and where he chose, traded with the Indians, sold his furs to whoever offered highest for them, dressed flauntingly, and generally had an Indian wife and half-breed children. They prided themselves on their hardihood and courage, even on their recklessness and profligacy. Each claimed to own the best horse, to have had the wildest adventures, to have made the most narrow escapes, to have killed the greatest number of bears and Indians, to be the greatest favorite with the Indian belles, the greatest consumer of alcohol, and to have the most money to spend, i.e. the largest credit on the books of the company. If his hearers did not believe him, he was ready to run a race with him, to beat him at *old sledge*, or to fight, if fighting was preferred—ready to prove what he affirmed in any manner the company pleased.

If the free trapper had a wife, she moved with the camp to which he attached himself, being furnished with a fine horse, caparisoned in the gayest and costliest manner. Her dress was of the finest goods the market afforded, and was suitably ornamented with beads, ribbons, fringes, and feathers. Her rank, too, as a free trapper's wife, gave her consequence not only in her own eyes, but in those of her tribe, and protected her from that slavish drudgery to which as the wife of an Indian hunter or warrior she would have been subject. The only authority which the free trapper acknowledged was that of his Indian spouse, who generally ruled in the lodge, however her lord blustered outside.

One of the free trapper's special delights was to take in hand the raw recruits, to gorge their wonder with his boastful tales, and to amuse himself with shocking his pupil's civilized notions of propriety. Joe Meek did not escape this sort of *breaking in*, and if it should appear in the course of this narrative that he proved an apt scholar, it will but illustrate a truth—that high spirits and fine talents tempt the tempter to win them over to his ranks. But Joe was not won over all at once. He beheld the beautiful spectacle of the encampment as it has been described, giving life and enchantment to the summer landscape, changed into a scene of the wildest carousal, going from bad to worse, until from harmless noise and bluster it came to fighting and loss of life. At this first rendezvous, he was shocked to behold the revolting exhibition of four trappers playing a game of cards with the dead body of a comrade for a card table! Such was the indifference to all the natural and ordinary emotions which these veterans of the wilderness cultivated in themselves, and inculcated in those who came under their influence. Scenes like this at first had the effect of bringing feelings of homesickness, while it inspired by contrast a sort of penitential and religious feeling also. According to Meek's account of those early days in the mountains, he said some secret prayers, and shed some secret tears. But this did not last long. The force of example, and especially the force of ridicule, is very potent with the young, nor are we quite free from their influence later in life.

If the gambling, swearing, drinking, and fighting at first astonished and alarmed the unsophisticated Joe, he found at the same time something to admire, and that he felt to be congenial with his own disposition, in the fearlessness, the contempt of sordid gain, the hearty merriment and frolicsome abandon of the better portion of the men about him. A spirit of emulation arose in him to become as brave as the bravest, as hardy as the hardiest, and as gay as the gayest, even while his feelings still revolted at many things which his heroic models were openly guilty of. If at any time in the future course of this narrative, Joe is discovered to have taken leave of his early scruples, the reader will considerately remember the associations by which he was surrounded for years, until the memory of the pious teachings of his childhood was nearly, if not quite, obliterated. To *nothing extenuate, nor set down aught in malice*, should be the frame of mind in which both the writer and reader of Joe's adventures should strive to maintain himself.

Before our hero is ushered upon the active scenes of a trapper's life, it may be well to present to the reader a sort of *guide to camp life*, in order that he may be able to understand some of its technicalities, as they may be casually mentioned hereafter.

When the large camp is on the march, it has a leader, generally one of the Booshways, who rides in advance, or at the head of the column. Near him is a lead mule, chosen for its qualities of speed and trustworthiness, on which are packed two small trunks that balance each other like panniers, and which contain the company's books, papers, and articles of agreement with the men. Then follow the pack animals, each one bearing three packs—one on each side, and one on top—so nicely adjusted as not to slip in traveling. These are in charge of certain men called camp-keepers, who have each three of these to look after. The trappers and hunters have two horses, or mules, one to ride, and one to pack their traps. If there are women and children in the train, all are mounted. Where the country is safe, the caravan moves in single file, often stretching out for half or three-quarters of a mile. At the end of the column rides the second man, or *little Booshway*, as the men call him, usually a hired officer, whose business it is to look after the order and condition of the whole camp.

On arriving at a suitable spot to make the night camp, the leader stops and dismounts in the particular space which is to be devoted to himself in its midst. The others, as they come up, form a circle, the *second man* bringing up the rear, to be sure all are there. He then proceeds to appoint every man a place in the circle, and to examine the horses' backs to see if any are sore. The horses are then turned out, under a guard, to graze, but before darkness comes on are placed inside the ring, and picketed by a stake driven in the earth, or with two feet so tied together as to prevent easy or free locomotion. The men are divided into messes: so many trappers and so many camp-keepers to a mess. The business of eating is not a very elaborate one, where the sole article of diet is meat, either dried or roasted. By a certain hour, all is quiet in camp, and only the guard is awake. At times during the night, the leader, or the officer of the guard, gives the guard a challenge—"All's well!" Which is answered by, "All's well!"

In the morning at daylight, or sometimes not till sunrise, according to the safe or dangerous locality, the second man comes forth from his lodge and cries in French, "*Leve, leve, leve, leve, leve*!" fifteen or twenty times, which is the command to rise. In about five minutes more, he cries out again, in French, "*Leche lego, leche lego*!" or turn out, turn out, at which command all come out from the lodges, and the horses are turned loose to feed, but not before a horseman has galloped all round the camp at some distance, and discovered everything to be safe in the neighborhood. Again, when the horses have been sufficiently fed, under the eye of a guard, they are driven up, the packs replaced, the train mounted, and once more, it moves off, in the order before mentioned.

In a settled camp, as in winter, there are other regulations. The leader and the second man occupy the same relative positions, but other minor regulations are observed. The duty of a trapper, for instance, in the trapping season, is only to trap and take care of his own horses. When he comes in at night, he takes his beaver to the clerk, and the number is counted off, and placed to his credit. Not he, but the camp-keepers, take off the skins and dry them. In the winter camp, there are six persons to a lodge: four trappers and two camp-keepers, therefore the trappers are well waited upon, their only duty being to hunt, in turns, for the camp. When a piece of game is brought in—a deer, an antelope, or buffalo meat—it is thrown down on the heap, which accumulates in front of the Booshway's lodge, and the second man stands by and cuts it up, or has it cut up for him. The first man who chances to come along, is ordered to stand still and turn his back to the pile of game, while the *little Booshway* lays hold of a piece that has been cut off, and asks in a loud voice—"who will have this?"—and the man answering for him, says, "the Booshway," or perhaps "number six," or "number twenty"—meaning certain messes, and the number is called to come and take their meat. In this blind way the meat is portioned off, strongly reminding one of the game of *button, button, who has the button?* In this chance game of the meat, the Booshway fares no better than his men, unless, in rare instances, the little Booshway should indicate to the man who calls off, that a certain choice piece is designed for the mess of the leader or the second man.

A gun is never allowed to be fired in camp under any provocation, short of an Indian raid, but the guns are frequently inspected, to see if they are in order, and woe to the careless camp-keeper who neglects this or any other duty. When the second man comes around, and finds a piece of work imperfectly done, whether it be cleaning the firearms, making a hair rope, or a skin lodge, or washing a horse's back, he does not threaten the offender with personal chastisement, but calls up another man and asks him, "Can *you* do this properly?"

"Yes, sir."

"I will give you ten dollars to do it." And the ten dollars is set down to the account of the inefficient camp-keeper. But he does not risk forfeiting another ten dollars in the same manner.

In the spring, when the camp breaks up, the skins that have been used all winter for lodges are cut up to make moccasins because from their having been thoroughly smoked by the lodge fires, they do not shrink in wetting, like raw skins. This is an important quality in a moccasin, as a trapper is almost constantly in the water, and should not his moccasins be smoked they will close upon his feet, in drying, like a vise. Sometimes, after trapping all day, the tired and soaked trapper lies down in his blankets at night, still wet. But by-and-by, he is wakened by the pinching of his moccasins, and is obliged to rise and seek the water again to relieve himself of the pain. For the same reason, when spring comes, the trapper is forced to cut off the lower half of his buckskin breeches and piece them down with blanket leggin's, which he wears all through the trapping season.

Such were a few of the peculiarities, and the hardships also, of a life in the Rocky Mountains. If the camp discipline and the dangers and hardships to which a raw recruit was exposed failed to harden him to the service in one year, he was rejected as a trifling fellow and sent back to the settlement the next year. It was not probable, therefore, that the mountain man often was detected in complaining at his lot. If he was miserable, he was laughed at, and he soon learned to laugh at his own miseries, as well as to laugh back at his comrades.

# 2

The business of the rendezvous occupied about a month. In this period, the men, Indian allies, and other Indian parties who usually visited the camp at this time, were all supplied with goods. The remaining merchandise was adjusted for the convenience of the different traders who should be sent out through all the country traversed by the company. Sublette then decided upon their routes, dividing up his forces into camps, which took each its appointed course, detaching as it proceeded small parties of trappers to all the hunting grounds in the neighborhood. These smaller camps were ordered to meet at certain times and places, to report progress, collect and cache their furs, and *count noses*. If certain parties failed to arrive, others were sent out in search for them.

This year, in the absence of Smith and Jackson, a considerable party was dispatched, under Milton Sublette, brother of the captain, and two other free trappers and traders, Frapp and Jervais, to traverse the country down along the Bighorn River.

Captain Sublette took a large party, among whom was Joe Meek, across the mountains to trap on the Snake River, in opposition to the Hudson's Bay Company. The Rocky Mountain Fur Company had hitherto avoided this country, except when Smith had once crossed to the headwaters of the Snake with a small party of five trappers. But Smith and Sublette had determined to oppose themselves to the British traders who occupied so large an extent of territory presumed to be American, and it had been agreed between them to meet this year on Snake River on Sublette's return from St. Louis, and Smith's from his California tour. What befell Smith's party before reaching the Columbia, has already been related, also his reception by the Hudson's Bay Company, and his departure from Vancouver.

Sublette led his company up the valley of the Wind River, across the mountains, and on to the very headwaters of the Lewis or Snake River. Here, he fell in with Jackson, in the valley of Lewis Lake, called Jackson's Hole, and remained on the borders of this lake for some time, waiting for Smith, whose non-appearance began to create a good deal of uneasiness. At length, runners were dispatched in all directions, looking for the lost Booshway.

The detachment to which Meek was assigned had the pleasure and honor of discovering the hiding place of the missing partner, which was in Pierre's Hole, a mountain valley about thirty miles long and of half that width, which subsequently was much frequented by the camps of the various fur companies. He was found trapping and exploring, in company with four men only, one of whom was Black, who with him escaped from the Umpqua Indians, as before related.

Notwithstanding the excitement and elation attendant upon

the success of his party, Meek found time to admire the magnificent scenery of the valley, which is bounded on two sides by broken and picturesque ranges, and overlooked by that magnificent group of mountains, called the Three Tetons, towering to a height of fourteen thousand feet. This emerald cup, set in its rim of amethystine mountains, was so pleasant a sight to the mountain men that camp was moved to it without delay, where it remained until sometime in September, recruiting its animals and preparing for the fall hunt.

Here again, the trappers indulged in their noisy sports and rejoicing, ostensibly on account of the return of the long-absent Booshway. There was little said of the men who had perished in that unfortunate expedition. "Poor fellow! out of luck," was the usual burial rite which the memory of a dead comrade received. So much and no more. They could indulge in noisy rejoicings over a lost comrade restored, but the dead one was not mentioned. Nor was this apparently heartless and heedless manner so irrational or unfeeling as it seemed. Everybody understood one thing in the mountains—that he must keep his life by his own courage and valor, or at the least by his own prudence. Unseen dangers always lay in wait for him. The arrow or tomahawk of the Indian, the blow of the grizzly bear, the misstep on the dizzy or slippery height, the rush of boiling and foaming floods, freezing cold, famine—these were the most common forms of peril, yet did not embrace even then all the forms in which Death sought his victims in the wilderness. The avoidance of painful reminders, such as the loss of a party of men, was a natural instinct, involving also a principle of self-defense—since to have weak hearts would be the surest road to defeat in the next dangerous encounter. To keep their hearts *big*, they must be gay, they must not remember the miserable fate of many of their one-time comrades. Think of that, stern moralist and martinet in propriety! Your fur collar hangs in the gas-lighted hall. In your luxurious dressing gown and slippers, by the warmth of a glowing grate, you muse upon the depravity of your fellow men. But imagine yourself, if you can, in the heart of an interminable wilderness. Let the snow be three or four feet deep, game scarce, Indians on your track: escaped from these dangers, once more beside a campfire, with a roast of buffalo meat on a stick before it, and several of your companions similarly escaped, and destined for the same chances to-morrow, around you. Do you fancy you should give much time to lamenting the less lucky fellows who were left behind frozen, starved, or scalped? Not you. You would be fortifying yourself against tomorrow when the same terrors might lay in wait for you. Jedediah Smith was a pious man, one of the few that ever resided in the Rocky Mountains, and led a band of reckless trappers, but he did not turn back to his camp when he saw it attacked on the Umpqua, nor stop to lament his murdered men. The law of self-preservation is strong in the wilderness. "Keep up your heart today, for tomorrow you may die," is the motto of the trapper.

In the conference which took place between Smith and Sublette, the former insisted that on account of the kind services of the Hudson's Bay Company toward himself and the three other survivors of his party, they should withdraw their trappers and traders from the western side of the mountains for the present, so as not to have them come in conflict with those of that company. To this proposition, Sublette reluctantly consented, and orders were issued for moving once more to the east, before going into winter camp, which was appointed for the Wind River Valley.

In the meantime, Joe Meek was sent out with a party to take his first hunt for beaver as a hired trapper. The detachment to which he belonged traveled down Pierre's Fork, the stream which watered the valley of Pierre's Hole, to its junction with Lewis's and Henry's Forks where they unite to form the great Snake River. While trapping in this locality, the party became aware of the vicinity of a roving band of Blackfeet, and in consequence, redoubled their usual precautions while on the march.

The Blackfeet were the tribe most dreaded in the Rocky Mountains, and went by the name of *Bugs Boys*, which rendered into good English, meant *the devil's own*. They are now so well known that to mention their characteristics seems like repeating a *twice-told tale*, but as they will appear so often in this narrative, Irving's account of them as he had it from Bonneville when he was fresh from the mountains, will, after all, not be out of place. "These savages," he says, "are the most dangerous banditti of the mountains, and the inveterate foe of the trapper. They are Ishmaelites of the first order, always with weapon in hand, ready for action. The young braves of the tribe, who are destitute of property, go to war for booty, to gain horses, and acquire the means of setting up a lodge, supporting a family, and entitling themselves to a seat in the public councils. The veteran warriors fight merely for the love of the thing, and the consequence which success gives them among their people. They are capital horsemen, and are generally well mounted on short, stout horses, similar to the prairie ponies, to be met with in St. Louis. When on a war party, however, they go on foot to enable them to skulk through the country with greater secrecy, to keep in thickets and ravines, and use more adroit subterfuges and stratagems. Their mode of warfare is entirely by ambush, surprise, and sudden assaults in the nighttime. If they succeed in causing a panic, they dash forward with headlong fury, if the enemy is on the alert, and shows no signs of fear, they become wary and deliberate in their movements.

"Some of them are armed in the primitive style, with bows and arrows, and the greater part have American fusees, made after the fashion of those of the Hudson's Bay Company. These they procure at the trading post of the American Fur Company on Maria's River, where they traffic their peltries for arms, ammunition, clothing, and trinkets. They are extremely fond of spirituous liquors and tobacco, for which nuisances they are ready to exchange, not merely their guns and horses but even their wives and daughters. As they are a treacherous race and have cherished a lurking hostility to the Whites ever since one of their tribe was killed by Mr. Lewis, the associate of General Clarke, in his exploring expedition across the Rocky Mountains, the American Fur Company is obliged constantly to keep at their post a garrison of sixty or seventy men.

"Under the general name of Blackfeet are comprehended several tribes, such as the Surcies, the Peagans, the Blood Indians, and the Gros Ventres of the Prairies, who roam about the Southern branches of the Yellowstone and Missouri Rivers, together with some other tribes further north. The bands infesting the Wind River Mountains, and the country adjacent, at the time of which we are treating, were Gros Ventres *of the Prairies,* which are not to be confounded with the Gros Ventres *of the Missouri*, who keep about the *lower* part of that river, and are friendly to the white men.

"This hostile band keeps about the headwaters of the Missouri, and numbers about nine hundred fighting men. Once in the course of two or three years they abandon their usual abodes and make a visit to the Arapahos of the Arkansas. Their route lies either through the Crow country, and the Black Hills, or through the lands of the Nez Percé, Flatheads, Bannacks, and Shoshones. As they enjoy their favorite state of hostility with all these tribes, their expeditions are prone to be conducted in the most lawless and predatory style, nor do they hesitate to extend their maraudings to any party of White men they meet with, following their trail, hovering about their camps, waylaying and dogging the caravans of the free traders, and murdering the solitary trapper. The consequences are frequent and desperate fights between them and the mountaineers, in the wild defiles and fastnesses of the Rocky Mountains." Such were the Blackfeet at the period of which we are writing, nor has their character changed at this day, as many of the Montana miners know to their cost.

# 3

## 1830

Sublette's camp commenced moving back to the east side of the Rocky Mountains in October. Its course was up Henry's Fork of the Snake River, through the North Pass to Missouri Lake, in which rises the Madison Fork of the Missouri River. The beaver were very plenty on Henry's Fork, and our young trapper had great success in making up his packs, having learned the art of setting his traps very readily. The manner in which the trapper takes his game is as follows:

He has an ordinary steel trap weighing five pounds, attached to a chain five feet long, with a swivel and ring at the end, which plays round what is called the *float*, a dry stick of wood, about six feet long. The trapper wades out into the stream, which is shallow, and cuts with his knife a bed for the trap, five or six inches under water. He then takes the float out the whole length of the chain in the direction of the centre of the stream, and drives it into the mud, so fast that the beaver cannot draw it out, and at the same time, tying the other end by a thong to the bank. A small stick or twig, dipped in musk or castor, serves for bait, and is placed so as to hang directly above the trap, which is now set. The trapper then throws water plentifully over the adjacent bank to conceal any footprints or scent by which the beaver would be alarmed, and going to some distance, wades out of the stream.

In setting a trap, several things are to be observed with care: first, that the trap is firmly fixed, and the proper distance from the bank—for if the beaver can get on shore with the trap, he will cut off his foot to escape and second, that the float is of dry wood, for should it not be, the little animal will cut it off at a stroke, and swimming with the trap to the middle of the dam, be drowned by its weight. In the latter case, when the hunter visits his traps in the morning, he is under the necessity of plunging into the water and swimming out to dive for the missing trap, and his game. Should the morning be frosty and chill, as it very frequently is in the mountains, diving for traps is not the pleasantest exercise. In placing the bait, care must be taken to fix it just where the beaver in reaching it will spring the trap. If the bait-stick be placed high, the hind foot of the beaver will be caught: if low, his forefoot.

The manner in which the beavers make their dam, and construct their lodge, has long been reckoned among the wonders of the animal creation, and while some observers have claimed for the little creature more sagacity than it really possesses, its instinct is still sufficiently wonderful. It is certainly true that it knows how to keep the water of a stream to a certain level, by means of an obstruction, and that it cuts down trees for the purpose of backing up the water by a dam. It is not true, however, that it can always fell a tree in the direction required for this purpose. The timber about a beaver dam is felled in all directions, but as trees that grow near the water, generally lean toward it, the tree, when cut, takes the proper direction by gravitation alone. The beaver then proceeds to cut up the fallen timber into lengths of about three feet, and to convey them to the spot where the dam is to be situated, securing them in their places by means of mud and stones. The work is commenced when the water is low, and carried on as it rises, until it has attained the desired height. And not only is it made of the requisite height and strength, but its shape is suited exactly to the nature of the stream in which it is built. If the water is sluggish the dam is straight, if rapid and turbulent, the barrier is constructed of a convex form, the better to resist the action of the water.

When the beavers have once commenced a dam, its extent and thickness are continually augmented, not only by their labors, but by accidental accumulations, thus accommodating itself to the size of the growing community. At length, after a lapse of many years, the water being spread over a considerable tract, and filled up by yearly accumulations of driftwood and earth, seeds take root in the new made ground, and the old beaver-dams become green meadows, or thickets of cottonwood and willow.

The food on which the beaver subsists is the bark of the young trees in its neighborhood, and when laying up a winter store, the whole community join in the labor of selecting, cutting up, and carrying the strips to their storehouses underwater. They do not, as some writers have affirmed, when cutting wood for a dam strip off the bark and store it in their lodges for winter consumption, but only carry underwater the stick with the bark on.

The beaver has two incisors and eight molars in each jaw, and empty hollows where the canine teeth might be. The upper pair of cutting teeth extend far into the jaw, with a curve of rather more than a semicircle, and the lower pair of incisors form rather less than a semicircle. Sometimes, one of these teeth gets broken and then the opposite tooth continues growing until it forms a nearly complete circle. The chewing muscle of the beaver is strengthened by tendons in such a way as to give it great power. But more is needed to enable the beaver to eat wood. The insalivation of the dry food is provided for by the extraordinary size of the salivary glands.

Now, every part of these instruments is of vital importance to the beavers. The loss of an incisor involves the formation of an obstructive circular tooth, and deficiency of saliva renders the food indigestible, and when old age comes and the enamel is worn down faster than it is renewed, the beaver is no longer able to cut branches for its support. Old, feeble and poor, unable to borrow, and ashamed to beg, he steals cuttings, and subjects himself to the penalty assigned to theft. Aged beavers are often found dead with gashes in their bodies, showing that they have been killed by their mates. In the fall of 1864, a very aged beaver was caught in one of the dams of the Esconawba River, and this was the reflection of a great authority on the occasion, one Ah-she-goes, an Ojibwa trapper: 'Had he escaped the trap he would have been killed before the winter was over, by other beavers, for stealing cuttings.'

When the beavers are about two or three years old, their

teeth are in their best condition for cutting. On the Upper Missouri, they cut the cotton tree and the willow bush, around Hudson's Bay and Lake Superior, in addition to the willow, they cut the poplar and maple, hemlock, spruce and pine. The cutting is round and round, and deepest upon the side on which they wish the tree to fall. Indians and trappers have seen beavers cutting trees. The felling of a tree is a family affair. No more than a single pair with two or three young ones are engaged at a time. The adults take the cutting in turns, one gnawing and the other watching, and occasionally a youngster trying his incisors. The beaver while gnawing sits on his plantigrade hind legs, which keep him conveniently upright. When the tree begins to crackle the beavers work cautiously, and when it crashes down, they plunge into the pond, fearful lest the noise should attract an enemy to the spot. After the tree-fall, comes the lopping of the branches. A single tree may be winter provision for a family. Branches five or six inches thick have to be cut into proper lengths for transport and are then taken home."

The lodge of a beaver is generally about six feet in diameter, on the inside, and about half as high. They are rounded or dome-shaped on the outside, with very thick walls, and communicate with the land by subterranean passages, below the depth at which the water freezes in winter. Each lodge is made to accommodate several inmates, who have their beds ranged round the walls, much as the Indian does in his tent. They are very cleanly, too, and after eating, carry out the sticks that have been stripped, and either use them in repairing their dam, or throw them into the stream below.

During the summer months the beavers abandon their lodges, and disport themselves about the streams, sometimes going on long journeys, or if any remain at home, they are the mothers of young families. About the last of August, the community returns to its home and begins preparations for the domestic cares of the long winter months.

An exception to this rule is that of certain individuals, who have no families, make no dam, and never live in lodges, but burrow in subterranean tunnels. They are always found to be males, whom the French trappers call *les parasseux*, or idlers, and the American trappers, *bachelors*. Several of them are sometimes found in one abode, which the trappers facetiously denominate *bachelor's hall*. Being taken with less difficulty than the more domestic beaver, the trapper is always glad to come upon their habitations.

The trapping season is usually in the spring and autumn. But should the hunters find it necessary to continue their work in winter, they capture the beaver by sounding on the ice until an aperture is discovered, when the ice is cut away and the opening closed up. Returning to the bank, they search for the subterranean passage, tracing its connection with the lodge, and by patient watching succeed in catching the beaver on some of its journeys between the water and the land. This, however, is not often resorted to when the hunt in the fall has been successful, or when not urged by famine to take the beaver for food.

"Occasionally it happens," says Captain Bonneville, "that several members of a beaver family are trapped in succession. The survivors then become extremely shy, and can scarcely be *brought to medicine*, to use the trappers' phrase for *taking the bait*. In such case, the trapper gives up the use of the bait, and conceals his traps in the usual paths and crossing places of the household. The beaver being now completely *up to trap*, approaches them cautiously, and springs them, ingeniously, with a stick. At other times, he turns the traps bottom upward, by the same means, and occasionally even drags them to the barrier, and conceals them in the mud. The trapper now gives up the contest of ingenuity, and shouldering his traps, marches off, admitting that he is not yet *up to beaver*."

Before the camp moved from the Forks of the Snake River, the haunting Blackfeet made their appearance openly. It was here that Meek had his first battle with that nation, with whom he subsequently had many a savage contest. They attacked the camp early in the morning, just as the call to turn out had sounded. But they had miscalculated their opportunity, the design having evidently been to stampede the horses and mules, at the hour and moment of their being turned loose to graze. They had been too hasty by a few minutes, so that when they charged on the camp pell-mell, firing a hundred guns at once, to frighten both horses and men, it happened that only a few of the animals had been turned out, and they had not yet got far off. The noise of the charge only turned them back to camp.

In an instant's time, Fitzpatrick was mounted, and commanding the men to follow, he galloped at headlong speed round and round the camp, to drive back such of the horses as were straying, or had been frightened from their pickets. In this race, two horses were shot under him, but he escaped, and the camp-horses were saved. The battle now was to punish the thieves. They took their position, as usual with Indian fighters, in a narrow ravine, from whence the camp was forced to dislodge them, at a great disadvantage. This they did do, at last, after six hours of hard fighting, in which a few men were wounded, but none killed. The thieves skulked off, through the canyon, when they found themselves defeated, and were seen no more until the camp came to the woods which cover the western slope of the Rocky Mountains.

But as the camp moved eastward, or rather in a northeasterly direction, through the pine forests between Pierre's Hole and the headwaters of the Missouri, it was continually harassed by Blackfeet, and required a strong guard at night, when these marauders delighted to make an attack. The weather by this time was very cold in the mountains, and chilled the marrow of our young Virginian. The travel was hard, too, and the recruits pretty well worn out.

One cold night, Meek was put on guard on the further side of the camp, with a veteran named Reese. But neither the veteran nor the youngster could resist the approaches of *tired nature's sweet restorer* and went to sleep at their post of duty. When, during the night, Sublette came out of his tent and gave the challenge, "All's well!" there was no reply. To quote Meek's own language, "Sublette came round the horse-pen swearing and snorting. He was powerful mad. Before he got to where Reese was, he made so much noise that he waked him, and Reese, in a loud whisper, called to him, 'Down, Billy! Indians!' Sublette got down on his belly mighty quick. 'Whar? whar?' he asked.

"'They were right there when you hollered so,' said Reese.

"'Where is Meek?' whispered Sublette.

"'He is trying to shoot one,' answered Reese, still in a whisper.

"Reese then crawled over to whar I war, and told me what had been said, and informed me what to do. In a few minutes I crept cautiously over to Reese's post, when Sublette asked me how many Indians had been thar, and I told him I couldn't make out their number. In the morning a pair of Indian moccasins war found whar Reese *saw the Indians*, which I had *taken care to leave there* and thus confirmed, our story got us the credit of vigilance, instead of our receiving our just dues for neglect of duty."

It was sometime during the fall hunt in the Pine Woods, on the west side of the Rocky Mountains, that Meek had one of his earliest adventures with a bear. Two comrades, Craig and Nelson, and himself, while out trapping, left their horses, and traveled up a creek on foot, in search of beaver. They had not proceeded any great distance, before they came suddenly face to face with a red bear, so suddenly, indeed, that the men made a spring for the nearest trees. Craig and Meek ascended a large pine, which chanced to be nearest, and having many limbs, was easy to climb. Nelson happened to take to one of two small trees that grew close together, and the bear, fixing upon him for a victim, undertook to climb after him. With his back against one of these small trees, and his feet against the other, his bearship succeeded in reaching a point not far below Nelson's perch, when the trees opened with his weight, and down he went, with a shock that fairly shook the ground. But this bad luck only seemed to infuriate the beast, and up he went again, with the same result, each time almost reaching his enemy. With the second tumble he was not the least discouraged, but started up the third time, only to be dashed once more to the ground when he had attained a certain height. At the third fall, however, he became thoroughly disgusted with his want of success, and turned and ran at full speed into the woods.

"Then," says Meek, "Craig began to sing, and I began to laugh, but Nelson took to swearing. 'O yes, you can laugh and sing now,' says Nelson, 'but you war quiet enough when the bear was around.' 'Why, Nelson,' I answered, 'you wouldn't have us noisy before that distinguished guest of yours?' But Nelson damned the wild beast, and Craig and I laughed, and said he didn't seem wild a bit. That's the way we hector each other in the mountains. If a man gets into trouble, he is only laughed at: 'let him keep out, let him have better luck,' is what we say."

The country traversed by Sublette in the fall of 1829, was unknown at that period, even to the fur companies, they having kept either farther to the south or to the north. Few, if any, white men had passed through it since Lewis and Clarke discovered the headwaters of the Missouri and the Snake Rivers[1], which flow from the opposite sides of the same mountain peaks. Even the toils and hardships of passing over mountains at this season of the year, did not deprive the trapper of the enjoyment of the magnificent scenery the region afforded. Splendid views, however, could not long beguile men who had little to eat, and who had yet a long journey to accomplish in cold, and surrounded by dangers, before reaching the wintering ground.

In November, the camp left Missouri Lake on the east side of the mountains, and crossed over, still northeasterly, on to the Gallatin Fork of the Missouri River, passing over a very rough and broken country. They were, in fact, still in the midst of mountains, being spurs of the great Rocky range, and equally high and rugged. A particularly high mountain lay between them and the main Yellowstone River. This they had just crossed, with great fatigue and difficulty, and were resting the camp and horses for a few days on the river's bank, when the

1. Lewis and Clark by no stretch of the imagination or of topography discovered the headwaters of the Snake River.

Blackfeet once more attacked them in considerable numbers. Two men were killed in this fight, and the camp thrown into confusion by the suddenness of the alarm. Capt. Sublette, however, got off, with most of his men, still pursued by the Indians.

Not so our Joe, who this time was not in luck, but was cut off from camp, alone, and had to flee to the high mountains overlooking the Yellowstone. Here was a situation for a nineteen-year-old raw recruit! Knowing that the Blackfeet were on the trail of the camp, it was death to proceed in that direction. Some other route must be taken to come up with them, the country was entirely unknown to him, the cold severe, his mule, blanket, and gun, his only earthly possessions. On the latter he depended for food, but game was scarce, and besides, he thought the sound of his gun would frighten himself, so alone in the wilderness, swarming with stealthy foes.

Hiding his mule in a thicket, he ascended to the mountaintop to take a view of the country, and decide upon his course. And what a scene was that for the miserable boy, whose chance of meeting with his comrades again was small indeed! At his feet rolled the Yellowstone River, coursing away through the great plain to the eastward. To the north, his eye follows the windings of the Missouri, as upon a map, but playing at hide-and-seek in among the mountains. Looking back, he saw the River Snake stretching its serpentine length through lava plains, far away, to its junction with the Columbia. To the north, and to the south, one white mountain rose above another as far as the eye could reach. What a mighty and magnificent world it seemed, to be alone in! Poor Joe succumbed to the influence of the thought, and wept.

Having indulged in this sole remaining luxury of life, Joe picked up his resolution, and decided upon his course. To the southeast lay the Crow country, a land of plenty—as the mountain man regards plenty—and there he could at least live, provided the Crows permitted him to do so. Besides, he had some hopes of falling in with one of the camps, by taking that course.

Descending the mountain to the hiding place of his mule, by which time it was dark night, hungry and freezing, Joe still could not light a fire, for fear of revealing his whereabouts to the Indians, nor could he remain to perish with cold. Travel he must, and travel he did, going he scarcely knew whither. Looking back upon the terrors and discomforts of that night, the veteran mountaineer yet regards it as about the most miserable one of his life. When day at length broke, he had made, as well as he could estimate the distance, about thirty miles. Traveling on toward the southeast, he had crossed the Yellowstone River, and still among the mountains, was obliged to abandon his mule and accouterments, retaining only one blanket and his gun. Neither the mule nor himself had broken fast in the last two days. Keeping a southerly course for twenty miles more, over a rough and elevated country, he came, on the evening of the third day, upon a band of mountain sheep. With what eagerness did he hasten to kill, cook, and eat! Three days of fasting was, for a novice, quite sufficient to provide him with an appetite.

Having eaten voraciously, and being quite overcome with fatigue, Joe fell asleep in his blanket, and slumbered quite deeply until morning. With the morning came biting blasts from the north, that made motion necessary if not pleasant. Refreshed by sleep and food, our traveler hastened on upon his solitary way, taking with him what sheep-meat he could carry, traversing the same rough and mountainous country as before. No incidents nor alarms varied the horrible and monotonous solitude of the wilderness. The very absence of anything to alarm was awful, for the bravest man is wretchedly nervous in the solitary presence of sublime nature. Even the veteran hunter of the mountains can never entirely divest himself of this feeling of awe, when his single soul comes face to face with God's wonderful and beautiful handiwork.

At the close of the fourth day, Joe made his lonely camp in a deep defile of the mountains, where a little fire and some roasted mutton again comforted his inner and outer man, and another night's sleep still farther refreshed his wearied frame. On the following morning, a very bleak and windy one, having breakfasted on his remaining piece of mutton, being desirous to learn something of the progress he had made, he ascended a low mountain in the neighborhood of his camp—and behold! the whole country beyond was smoking with the vapor from boiling springs, and burning with gasses, issuing from small craters, each of which was emitting a sharp whistling sound.

When the first surprise of this astonishing scene had passed, Joe began to admire its effect in an artistic point of view. The morning being clear, with a sharp frost, he thought himself reminded of the city of Pittsburg, as he had beheld it on a winter morning, a couple of years before. This, however, related only to the rising smoke and vapor, for the extent of the volcanic region was immense, reaching far out of sight. The general face of the country was smooth and rolling, being a level plain, dotted with cone-shaped mounds. On the summits of these mounds were small craters from four to eight feet in diameter. Interspersed among these, on the level plain, were larger craters, some of them from four to six miles across. Out of these craters issued blue flames and molten brimstone.

For some minutes, Joe gazed and wondered. Curious thoughts came into his head, about hell and the day of doom. With that natural tendency to reckless gaiety and humorous absurdities which some temperaments are sensible of in times of great excitement, he began to soliloquize. Said he, to himself, "I have been told the sun would be blown out, and the earth burned up. If this infernal wind keeps up, I shouldn't be surprised if the sun war blown out. If the earth is *not* burning up over thar, then it is that place the old Methodist preacher used to threaten me with. Anyway, it suits me to go and see what it's like."

On descending to the plain described, the earth was found to have a hollow sound, and seemed threatening to break through. But Joe found the warmth of the place most delightful, after the freezing cold of the mountains, and remarked to himself again that, "if it war hell, it war a more agreeable climate than he had been in for some time."

He had thought the country entirely desolate, as not a living creature had been seen in the vicinity, but while he stood gazing about him in curious amazement, he was startled by the report of two guns, followed by the Indian yell. While making rapid preparations for defense and flight, if either or both should be necessary, a familiar voice greeted him with the exclamation, "It *is* old Joe!" When the adjective *old* is applied to one of Meek's age at that time, it is generally understood to be a term of endearment. "My feelings you may imagine," says the *old Uncle Joe* of the present time, in recalling the adventure.

Being joined by these two associates, who had been looking for him, our traveler, no longer simply a raw recruit, but a hero of wonderful adventures, as well as the rest of the men, proceeded with them to camp, which they overtook the third day, attempting to cross the high mountains between the Yellowstone and the Bighorn Rivers. If Meek had seen hard times in the mountains alone, he did not find them much improved in camp. The snow was so deep that the men had to keep in advance, and break the road for the animals, and to make their condition still more trying, there were no provisions in camp, nor any prospect of plenty, for men or animals, until they should reach the buffalo country beyond the mountains.

During this scarcity of provisions, some of those amusing incidents took place with which the mountaineer will contrive to lighten his own and his comrades' spirits, even in periods of the greatest suffering. One which we have permission to relate, has reference to what Joe Meek calls the *meanest act of his life*.

While the men were starving, a negro boy, belonging to Jedediah Smith, by some means was so fortunate as to have caught a porcupine, which he was roasting before the fire. Happening to turn his back for a moment, to observe something in camp, Meek and Reese snatched the tempting viand and made off with it, before the darkey discovered his loss. But when it was discovered, what a wail went up for the embezzled porcupine! Suspicion fixed upon the guilty parties, but as no one would 'peach on white men to save a *nigger's* rights, the poor, disappointed boy could do nothing but lament in vain, to the great amusement of the men, who upon the principle that *misery loves company*, rather chuckled over than condemned. Meek's *mean act*.

There was a sequel, however, to this little story. So much did the negro dwell upon the event, and the heartlessness of the men toward him, that in the following summer, when Smith was in St. Louis, he gave the boy his freedom and two hundred dollars, and left him in that city, so that it became a saying in the mountains that, "the nigger got his freedom for a porcupine."

During this same march, a similar joke was played upon one of the men named Craig. He had caught a rabbit and put it up to roast before the fire—a tempting-looking morsel to starving mountaineers. Some of his associates determined to see how it tasted, and Craig was told that the Booshways wished to speak with him at their lodge. While he obeyed this supposed command, the rabbit was spirited away, never more to be seen by mortal man. When Craig returned to the campfire, and beheld the place vacant where a rabbit so late was nicely roasting, his passion knew no bounds, and he declared his intention of cutting it out of the stomach that contained it. But as finding the identical stomach which contained it involved the cutting open of many that probably did not, in the search, he was fain to relinquish that mode of vengeance, together with his hopes of a supper. As Craig is still living, and is tormented by the belief that he knows the man who stole his rabbit, Mr. Meek takes this opportunity of assuring him, upon the word of a gentleman, that *he* is not the man.

While on the march over these mountains, owing to the depth of the snow, the company lost a hundred head of horses and mules, which sank in the yet unfrozen drifts, and could not be extricated. In despair at their situation, Jedediah Smith one day sent a man named Harris to the top of a high peak to take a view of the country, and ascertain their position. After a toilsome scramble, the scout returned.

"Well, what did you see, Harris?" asked Smith anxiously.

"I saw the city of St. Louis, and one fellow taking a drink!" replied Harris, prefacing the assertion with a shocking oath.

Smith asked no more questions. He understood by the man's answer that he had made no pleasing discoveries, and knew that they had still a weary way before them to reach the plains below. Besides, Smith was a religious man, and the coarse profanity of the mountaineers was very distasteful to him. "A very mild man, and a Christian, and there were very few of them in the mountains," is the account given of him by the mountaineers themselves.

The camp finally arrived without loss of life, except to the animals, on the plains of the Bighorn River, and came upon the waters of the Stinking Fork, a branch of this river, which derives its unfortunate appellation from the fact that it flows through a volcanic tract similar to the one discovered by Meek on the Yellowstone plains. This place afforded as much food for wonder to the whole camp, as the former one had to Joe, and the men unanimously pronounced it the *back door to that country which divines preach about*. As this volcanic district had previously been seen by one of Lewis and Clarke's men, named Colter, while on a solitary hunt, and by him also denominated *hell*, there must certainly have been something very suggestive in its appearance.

If the mountains had proven barren, and inhospitably cold, this hot and sulfurous country offered no greater hospitality. In fact, the fumes which pervaded the air rendered it exceedingly noxious to every living thing, and the camp was fain to push on to the mainstream of the Bighorn River. Here signs of trappers became apparent, and spies having been sent out discovered a camp of about forty men, under Milton Sublette, brother of Captain William Sublette, the same that had been detached the previous summer to hunt in that country. Smith and Sublette then cached their furs, and moving up the river joined the camp of M. Sublette.

The manner of caching furs is this: A pit is dug to a depth of five or six feet in which to stand. The men then drift from this under a bank of solid earth, and excavate a room of considerable dimensions, in which the furs are deposited, and the apartment closed up. The pit is then filled up with earth, and the traces of digging obliterated or concealed. These caches are the only store-houses of the wilderness.

While the men were recruiting themselves in the joint camp, the alarm of "Indians!" was given, and hurried cries of "shoot! shoot!" were uttered on the instant. Captain Sublette, however, checked this precipitation, and ordering the men to hold, allowed the Indians to approach, making signs of peace. They proved to be a war party of Crows, who, after smoking the pipe of peace with the captain, received from him a present of some tobacco, and departed.

As soon as the camp was sufficiently recruited for traveling, the united companies set out again toward the south, and crossed the Horn mountains once more into Wind River Valley, having had altogether, a successful fall hunt[2], and made some

---

2. The route of this fall hunt of 1829, led by Sublette and Smith, is up the Henry's Fork of the Snake—though Newell says up the Lewis Fork—across to the headwaters of the Madison Fork of the Missouri, across to the Gallatin Fork, across to the Yellowstone River, across the Absaroka Mountains to the Stinking Water—the modern Shoshone—River, and down that to its sulfurous springs. In modern terms they went from near the present Idaho Falls, Idaho, north, then east across northern Yellowstone Park and on east to where Cody, Wyoming now

important explorations, notwithstanding the severity of the weather and the difficulty of mountain traveling. It was about Christmas when the camp arrived on Wind River, and the cold intense. While the men celebrated Christmas, as best they might under the circumstances, Capt. Sublette started to St. Louis with one man, Harris, called among mountain men Black Harris, on snowshoes, with a train of pack-dogs. Such was the indomitable energy and courage of this famous leader!

---

is. Then they descended the Stinking Water to the Big Horn River, made a cache, and went back across the Absaroka Mountains into the Wind River Valley. Mrs. Victor refers to this last crossing as of the *Horn Mountains*, which must mean the range dividing the valleys of the Big Horn and Wind rivers, and Bonneville's map —reproduced in Irving—shows that range marked Little Horn Mountains, they were later called the Yellowstone Mountains and are now called the Absarokas. The two most likely routes across them from the Big Horn River near the mouth of the Shoshone are back up the Shoshone and up its South Fork to the area of modern Dubois, or up the Big Horn and then up Owl Creek and over via its south fork to near Dubois. One of these routes must have been the one taken by John Colter on his long winter journey in 1807-08.

# 4

## 1830

The furs collected by Jackson's company were cached on the Wind River, and the cold still being very severe, and game scarce, the two remaining leaders, Smith and Jackson, set out on the first of January with the whole camp, for the buffalo country, on the Powder River, a distance of about one hundred and fifty miles. "Times were hard in camp," when mountains had to be crossed in the depth of winter.

The animals had to be subsisted on the bark of the sweet cottonwood[1], which grows along the streams and in the valleys on the east side of the Rocky Mountains, but is nowhere to be found west of that range. This way of providing for his horses and mules involved no trifling amount of labor, when each man

1. This assertion that cottonwood with bark edible for horses does not exist west of the Continental Divide is bizarre.

had to furnish food for several of them. To collect this bark, the men carried the smooth limbs of the cottonwood to camp, where, beside the campfire, they shaved off the sweet, green bark with a hunting-knife transformed into a drawing-knife by fastening a piece of wood to its point, or in case the cottonwood was not convenient, the bark was peeled off, and carried to camp in a blanket. So nutritious is it, that animals fatten upon it quite as well as upon oats.

In the large cottonwood bottoms on the Yellowstone River, it sometimes became necessary to station a double guard to keep the buffalo out of camp, so numerous were they, when the severity of the cold drove them from the prairies to these cottonwood thickets for subsistence. It was, therefore, of double importance to make the winter camp where the cottonwood was plenty since not only did it furnish the animals of the camp with food, but by attracting buffalo, made game plenty for the men. To such a hunter's paradise on Powder River, the camp was now traveling, and arrived, after a hard, cold march, about the middle of January, when the whole encampment went into winter quarters[2], to remain until the opening of spring.

This was the occasion when the mountain man *lived fat* and enjoyed life: a season of plenty, of relaxation, of amusement, of acquaintanceship with all the company, of gaiety, and of *busy idleness*. Through the day, hunting parties were coming and

2. The route of this winter move seems to be unknown. Surely Smith went down Wind River, left it at its northward turn, and crossed the plains south of the Big Horn Mountains to the headwaters of Powder River. The more direct route would have involved the crossing of the Absaroka and Big Horn Mountains in January of what seems, by Meek's recollection, to have been a winter of early snow accumulation.

going, and men were cooking, drying meat, making moccasins, cleaning their arms, wrestling, playing games, and in short, everything that an isolated community of hardy men could resort to for occupation, was resorted to by these mountaineers. Nor was there wanting, in the appearance of the camp, the variety, and that picturesque air imparted by a mingling of the native element, for what with their Indian allies, their native wives, and numerous children, the mountaineers' camp was a motley assemblage, and the trappers themselves, with their affectation of Indian coxcombry, not the least picturesque individuals.

The change wrought in a wilderness landscape by the arrival of the grand camp was wonderful indeed. Instead of nature's superb silence and majestic loneliness, there was the sound of men's voices in boisterous laughter, or the busy hum of conversation, the loud-resounding stroke of the axe, the sharp report of the rifle, the neighing of horses, and braying of mules, the Indian whoop and yell, and all that not unpleasing confusion of sound which accompanies the movements of the creature man. Over the plain, only dotted until now with shadows of clouds, or the transitory passage of the deer, the antelope, or the bear, were scattered hundreds of lodges and immense herds of grazing animals. Even the atmosphere itself seemed changed from its original purity, and became clouded with the smoke from many camp-fires. And all this change might go as quickly as it came. The tent struck and the march resumed, solitude reigned once more, and only the cloud dotted the silent landscape.

If the day was busy and gleesome, the night had its charms as well. Gathered about the shining fires, groups of men in fantastic costumes told tales of marvelous adventures, or sung some old-remembered song, or were absorbed in games of chance. Some of the better-educated men, who had once known and loved books, but whom some mishap in life had banished to the wilderness, recalled their favorite authors, and recited passages once treasured, now growing unfamiliar, or whispered to some chosen confrere the saddened history of his earlier years, and charged him thus and thus, should ever-ready death surprise himself in the next spring's hunt.

It will not be thought discreditable to our young trapper, Joe, that he learned to read by the light of the campfire. Becoming sensible, even in the wilderness, of the deficiencies of his early education, he found a teacher in a comrade, named Green, and soon acquired sufficient knowledge to enjoy an old copy of Shakspeare, which, with a Bible, was carried about with the property of the camp.

In this life of careless gaiety and plenty, the whole company was allowed to remain without interruption, until the first of April, when it was divided, and once more started on the march. Jackson, or *Davey*, as he was called by the men, with about half the company, left for the Snake country. The remainder, among whom was Meek, started north, with Smith for commander, and James Bridger as pilot.[3]

Crossing the mountains, ranges of which divide the tributary

---

3. The route of the spring hunt of Jedediah Smith's brigade, 1830, was thus: From some point on Powder River west to the Tongue River and to Bovey's Fork of the Big Horn and still west over the Pryor Mountains to Clark's Fork, the Rosebud River, and north to the Yellowstone, probably near the present Big Timber, Montana—according to Morgan. Then north to the headwaters of the Musselshell River, down it, and on as far north as the Judith River. They returned via a presumedly similar route as far as the Big Horn, which they followed up to the Popo Agie, apparently going through Wind River Canyon, where the Big Horn changes its name to Wind River.

streams of the Yellowstone from each other, the first halt was made on Tongue River. From thence, the camp proceeded to the Bighorn River. Through all this country game was in abundance —buffalo, elk, and bear, and beaver also' plenty. In mountain phrase, "times were good on this hunt," beaver packs increased in number, and both men and animals were in excellent condition.

A large party usually hunted out the beaver and frightened away the game in a few weeks, or days, from any one locality. When this happened, the camp moved on. Should not game be plenty, it kept constantly on the move, the hunters and trappers seldom remaining out more than a day or two. Should the country be considered dangerous on account of Indians, it was the habit of the men to return every night to the encampment.

It was the design of Smith to take his command into the Blackfoot country, a region abounding in the riches which he sought, could they only be secured without coming into too frequent conflict with the natives: always a doubtful question concerning these savages. He had proceeded in this direction as far as Bovey's Fork of the Bighorn, when the camp was overtaken by a heavy fall of snow, which made traveling extremely difficult, and which, when melted, caused a sudden great rise in the mountain streams. In attempting to cross Bovey's Fork during the high water, he had thirty horses swept away, with three hundred traps: a serious loss in the business of hunting beaver.

In the manner described, pushing on through an unknown country, hunting and trapping as they moved, the company proceeded, passing another low chain of mountains, through a pass called Pryor's Gap, to Clark's Fork of the Yellowstone, thence to Rosebud River, and finally to the main Yellowstone River, where it makes a great bend to the east, enclosing a large plain covered with grass, and having also extensive cottonwood bottoms, which subsequently became a favorite wintering ground of the fur companies.

It was while trapping up in this country, on the Rosebud River, that an amusing adventure befell our trapper Joe. Being out with two other trappers, at some distance from the great camp, they had killed and supped off a fat buffalo cow. The night was snowy, and their camp was made in a grove of young aspens. Having feasted themselves, the remaining store of choice pieces was divided between, and placed, hunter fashion, under the heads of the party, on their betaking themselves to their blanket couches for the night. Neither Indian nor wild beast disturbed their repose, as they slept, with their guns beside them, filled with comfort and plenty. But who ever dreams of the presence of a foe under such circumstances? Certainly not our young trapper, who was only awakened about day-break by something very large and heavy walking over him, and snuffing about him with a most insulting freedom. It did not need Yankee powers of guessing to make out who the intruder in camp might be: in truth, it was only too disagreeably certain that it was a full-sized grizzly bear, whose keenness of smell had revealed to him the presence of fat cow-meat in that neighborhood.

"You may be sure," says Joe, "that I kept very quiet, while that bar helped himself to some of my buffalo meat, and went a little way off to eat it. But Mark Head, one of the men, raised up, and back came the bar. Down went our heads under the blankets, and I kept mine covered pretty snug, while the beast took another walk over the bed, but finally went off again to a little distance. Mitchel then wanted to shoot, but I said, 'no, no, hold on, or the brute will kill us, sure.' When the bar heard our voices, back he ran again, and jumped on the bed as before. I'd have been happy to have felt myself sinking ten feet underground, while that bar promenaded over and around us! However, he couldn't quite make out our style, and finally took fright, and ran off down the mountain. Wanting to be revenged for his impudence, I went after him, and seeing a good chance, shot him dead. Then I took my turn at running over him a while!"

Such are the not infrequent incidents of the trapper's life, which furnish him with material, needing little embellishment to convert it into those wild tales with which the nights are whiled away around the winter campfire.

Arrived at the Yellowstone with his company, Smith found it necessary, on account of the high water, to construct Bull-boats for the crossing. These are made by stitching together buffalo hides, stretching them over light frames, and paying the seams with elk tallow and ashes. In these light wherries, the goods and people were ferried over, while the horses and mules were crossed by swimming.

The mode usually adopted in crossing large rivers, was to spread the lodges on the ground, throwing on them the light articles, saddles, etc. A rope was then run through the pin-holes around the edge of each, when it could be drawn up like a reticule. It was then filled with the heavier camp goods, and being tightly drawn up, formed a perfect ball. A rope being tied to it, it was launched on the water, the children of the camp on top, and the women swimming after and clinging to it, while a man, who had the rope in his hand, swam ahead holding on to his horse's mane. In this way, dancing like a cork on the waves, the lodge was piloted across, and passengers as well as freight consigned, undamaged, to the opposite shore. A large camp of three hundred men, and one hundred women and children were frequently thus crossed in one hour's time.

The camp was now in the excellent but inhospitable country of the Blackfeet, and the commander redoubled his precautions, moving on all the while to the Mussel Shell, and thence to the Judith River. Beaver were plenty and game abundant, but the vicinity of the large village of the Blackfeet made trapping impracticable. Their war upon the trappers was ceaseless. Their thefts of traps and horses ever recurring: and Smith, finding that to remain was to be involved in incessant warfare, without hope of victory or gain, at length gave the command to turn back, which was cheerfully obeyed: for the trappers had been very successful on the spring hunt, and thinking discretion some part at least of valor, were glad to get safe out of the Blackfoot country with their rich harvest of beaver skins.

The return march was by the way of Pryor's Gap, and up the Bighorn, to Wind River, where the cache was made in the previous December. The furs were now taken out and pressed, ready for transportation across the plains. A party was also dispatched, under Mr. Tullock, to raise the cache on the Bighorn River. Among this party was Meek, and a Frenchman named Ponto. While digging to come at the fur, the bank above caved in, falling upon Meek and Ponto, killing the latter almost instantly. Meek, though severely hurt, was taken out alive: while poor Ponto was *rolled in a blanket, and pitched into the river*. So rude were the burial services of the trapper of the Rocky Mountains.

Meek was packed back to camp, along with the furs, where he soon recovered. Sublette arrived from St. Louis with fourteen wagons loaded with merchandise, and two hundred additional men for the service. Jackson also arrived from the Snake country with plenty of beaver, and the business of the yearly rendezvous began.[4] Then, the scenes previously described were re-enacted. Beaver, the currency of the mountains, was plenty that year, and goods were high accordingly. A thousand dollars a day was not too much for some of the most reckless to spend on their squaws, horses, alcohol, and themselves. For *alcohol* was the beverage of the mountaineers. Liquors could not be furnished to the men in that country. Pure alcohol was what they *got tight on*, and a desperate tight it was, to be sure!

An important change took place in the affairs of the Rocky Mountain Company at this rendezvous. The three partners, Smith, Sublette, and Jackson, sold out to a new firm, consisting of Milton Sublette, James Bridger, Fitzpatrick, Frapp, and Jervais, with the new company retaining the same name and style as the old.

The old partners left for St. Louis, with a company of seventy men, to convoy the furs. Two of them never returned to the Rocky Mountains, and one of them, Smith, being killed the following year, as will hereafter be related, and Jackson remaining in St. Louis, where, like a true mountain man, he dissipated his large and hard-earned fortune in a few years. Captain Sublette, however, continued to make his annual trips to and from the mountains for a number of years, and until the consolidation of another wealthy company with the Rocky

---

4. This rendezvous of 1830 was near the mouth of the Popo Agie River, close to the present Riverton, Wyoming. The firm that bought out Smith, Jackson & Sublette was called the Rocky Mountain Fur Company for the first time.

Mountain Company, continued to furnish goods to the latter, at a profit on St. Louis prices. His capital and experience enabled him to keep the new firm under his control to a large degree.

# 5

## 1830

The whole country lying upon the Yellowstone and its tributaries, and about the headwaters of the Missouri, at the time of which we are writing, abounded not only in beaver, but in buffalo, bear, elk, antelope, and many smaller kinds of game. Indeed, the buffalo used then to cross the mountains into the valleys about the headwaters of the Snake and Colorado Rivers, in such numbers that at certain seasons of the year, the plains and river bottoms swarmed with them. Since that day they have quite disappeared from the western slope of the Rocky Mountains, and are no longer seen in the same numbers on the eastern side.

Bear, although they did not go in herds, were rather uncomfortably numerous, and sometimes put the trapper to considerable trouble, and fright also, for very few were brave enough to willingly encounter the formidable grizzly, one blow of whose terrible paw, aimed generally at the hunter's head, if not arrested, lays him senseless and torn, an easy victim to the wrathful monster. A gunshot wound, if not directed with certainty to some vulnerable point, has only the effect to infuriate the beast, and make him trebly dangerous. From the fact that the bear always bites his wound, and commences to run with his head thus brought in the direction from which the ball comes, he is pretty likely to make a straight wake toward his enemy, whether voluntarily or not, and woe be to the hunter who is not prepared for him, with a shot for his eye, or the spot just behind the ear, where certain death enters.

In the frequent encounters of the mountain men with these huge beasts, many acts of wonderful bravery were performed, while some tragedies, and not a few comedies were enacted.

From something humorous in Joe Meek's organization, or some wonderful *luck* to which he was born, or both, the greater part of his adventures with bears, as with men, were of a humorous complexion, enabling him not only to have a story to tell, but one at which his companions were bound to laugh. One of these, which happened during the fall hunt of 1830, we will let him tell for himself:

"The first fall on the Yellowstone, Hawkins and myself were coming up the river in search of camp, when we discovered a very large bar on the opposite bank. We shot across, and thought we had killed him, fur he laid quite still. As we wanted to take some trophy of our victory to camp, we tied our mules and left our guns, clothes, and everything except our knives and belts, and swum over to whar the bar war. But instead of being dead, as we expected, he sprung up as we come near him, and took after us. Then you ought to have seen two naked men run! It war a race for life, and a close one, too. But we made the river first. The bank war about fifteen feet high above the water, and the river ten or twelve feet deep, but we didn't halt. Overboard we went, the bar after us, and in the stream about as quick as we war. The current war very strong, and the bar war about halfway between Hawkins and me. Hawkins was trying to swim downstream faster than the current war carrying the bar, and I war a trying to hold back. You can reckon that I swam! Every moment I felt myself being washed into the yawning jaws of the mighty beast, whose head war up the stream, and his eyes on me. But the current war too strong for him, and swept him along as fast as it did me. All this time, not a long one, we war looking for some place to land where the bar could not overtake us. Hawkins war the first to make the shore, unknown to the bar, whose head war still upstream, and he set up such a whooping and yelling that the bar landed too, but on the opposite side. I made haste to follow Hawkins, who had landed on the side of the river we started from, either by design or good luck: and then we traveled back a mile and more to whar our mules war left—a bar on one side of the river, and *two bares* on the other!"

Notwithstanding that a necessary discipline was observed and maintained in the fur traders' camp, there was at the same time a freedom of manner between the Booshways and the men, both hired and free, which could not obtain in a purely military organization, nor even in the higher walks of civilized life in cities. In the mountain community, motley as it was, as in other communities more refined, were some men who enjoyed almost unlimited freedom of speech and action, and others who were the butt of everybody's ridicule or censure. The leaders themselves did not escape the critical judgment of the men, and the estimation in which they were held could be inferred from the manner in which they designated them. Captain Sublette, whose energy, courage, and kindness entitled him to the admiration of the mountaineers, went by the name of *Billy:* his partner Jackson, was called *Davey.* Bridger, *old Gabe*, and so on. In the same manner, the men distinguished favorites or oddities among themselves, and to have the adjective *old* prefixed to a man's name signified nothing concerning his age, but rather that he was an object of distinction, though it did not always indicate, except by the tone in which it was pronounced, whether that distinction was an enviable one or not.

Whenever a trapper could get hold of any sort of story reflecting on the courage of a leader, he was sure at some time to make him aware of it, and these anecdotes were sometimes sharp answers in the mouths of careless camp-keepers. Bridger was once waylaid by Blackfeet, who shot at him, hitting his horse in several places. The wounds caused the animal to rear and pitch, by reason of which violent movements Bridger dropped his gun, and the Indians snatched it up, after which there was nothing to do except to run, which Bridger accordingly did. Not long after this, as was customary, the leader was making a circuit of the camp examining the camp-keeper's guns, to see if they were in order, and found that of one Maloney, an Irishman, in a very dirty condition.

"What would you do," asked Bridger, "with a gun like that, if the Indians were to charge on the camp?"

"Be Jasus, I would throw it to them, and run the way ye did," answered Maloney, quickly. It was sometime after this incident before Bridger again examined Maloney's gun.

A laughable story in this way went the rounds of the camp in this fall of 1830.[1] Milton Sublette was out on a hunt with Meek after buffalo, and they were just approaching the band on foot, at a distance apart of about fifty yards, when a large grizzly bear came out of a thicket and made after Sublette, who, when he perceived the creature, ran for the nearest cottonwood tree. Meek in the meantime, seeing that Sublette was not likely to escape, had taken sure aim, and fired at the bear, fortunately killing him. On running up to the spot where it laid, Sublette was discovered sitting at the foot of a cottonwood, with his legs and arms clasped tightly around it.

"Do you always climb a tree in that way?" asked Meek.

"I reckon you took the wrong end of it, that time, Milton!"

"I'll be damned, Meek, if I didn't think I was twenty feet up that tree when you shot," answered the frightened Booshway, and from that time the men never tired of alluding to Milton's manner of climbing a tree.

These were some of the mirthful incidents which gave occasion for a gaiety which had to be substituted for happiness, in the checkered life of the trapper, and there were like to be many

---

1. Bridger, Fitzpatrick, and Milton Sublette—three of the five new partners—led a brigade of close to a hundred men in this fall hunt, 1830. The route was down the Wind and Big Horn rivers, across to Clark's Fork, and across to the Yellowstone—probably around the present Big Timber-Livingston, Montana area—across to Smith River and down it to the Missouri near the Great Falls. Then they turned upstream and trapped to the Three Forks. From this point Meek has them going to Big Blackfoot River and down the west side of the mountains to Ogden's Hole and becoming embroiled with a Hudson's Bay brigade led by Peter Skene Ogden. If Joe isn't making this story up, and he likely is not, it cannot have happened at this time, because Ogden was on an expedition through Nevada to the Colorado River. Newell and others say more credibly that from the Three Forks the brigade crossed to the Yellowstone and went into winter quarters there. Early in the spring the brigade moved to Powder River, and there Meek's account rejoins it.

such, where there were two hundred men, each almost daily in the way of adventures by flood or field.

On the change in the management of the Company which occurred at the rendezvous this year, three of the new partners, Fitzpatrick, Sublette, and Bridger, conducted a large party, numbering over two hundred, from the Wind River to the Yellowstone, crossing thence to Smith's River, the Falls of the Missouri, three forks of the Missouri, and to the Big Blackfoot River. The hunt proved very successful, beaver were plentiful, and the Blackfeet shy of so large a traveling party. Although so long in their country, there were only four men killed out of the whole company during this autumn.

From the Blackfoot River the company proceeded down the west side of the mountains to the Forks of the Snake River, and after trapping for a short time in this locality, continued their march southward as far as Ogden's Hole, a small valley among the Bear River Mountains.

At this place they fell in with a trading and trapping party, under Mr. Peter Skeen Ogden, of the Hudson's Bay Company. And now commenced that irritating and reprehensible style of rivalry with which the different companies were accustomed to annoy one another. Accompanying Mr. Ogden's trading party were a party of Rockway Indians, who were from the north, and who were employed by the Hudson's Bay Company, as the Iroquois and Crows were, to trap for them. Fitzpatrick and associates camped in the neighborhood of Ogden's company, and immediately set about endeavoring to purchase from the Rockways and others, the furs collected for Mr. Ogden. Not succeeding by fair means, if the means to such an end could be called fair—they opened a keg of whiskey, which, when the

Indians had got a taste, soon drew them away from the Hudson's Bay trader, the regulations of whose company forbade the selling or giving of liquors to the Indians. Under its influence, the furs were disposed of to the Rocky Mountain Company, who in this manner, obtained nearly the whole product of their year's hunt. This course of conduct was naturally exceedingly disagreeable to Mr. Ogden, as well as unprofitable also, and a feeling of hostility grew up and increased between the two camps.

While matters were in this position, a stampede one day occurred among the horses in Ogden's camp, and two or three of the animals ran away and ran into the camp of the rival company. Among them was the horse of Mr. Ogden's Indian wife, which had escaped, with her babe hanging to the saddle.

Not many minutes elapsed before the mother, following her child and horse, entered the camp, passing right through it, and catching the now halting steed by the bridle. At the same moment, she espied one of her company's packhorses, loaded with beaver, which had also run into the enemy's camp. The men had already begun to exult over the circumstance, considering this chance load of beaver as theirs, by the laws of war. But not so the Indian woman. Mounting her own horse, she fearlessly seized the packhorse by the halter, and led it out of camp, with its costly burden.

At this undaunted action, some of the baser sort of men cried out, "Shoot her, shoot her!" But a majority interfered, with opposing cries of, "Let her go, let her alone, she's a brave woman: I glory in her pluck," and other like admiring expressions. While the clamor continued, the wife of Ogden had galloped away, with her baby and her packhorse.

As the season advanced, Fitzpatrick, with his other partners, returned to the east side of the mountains, and went into winter quarters on Powder River. In this trapper's *land of Canaan* they remained between two and three months. The other two partners, Frapp and Jervais, who were trapping far to the south, did not return until the following year.

While wintering it became necessary to send a dispatch to St. Louis on the company's business.[2] Meek and a Frenchman named Legarde, were chosen for this service, which was one of trust and peril also. They proceeded without accident, however, until the Pawnee villages were reached, when Legarde was taken prisoner. Meek, more cautious, escaped, and proceeded alone a few days' travel beyond, when he fell in with an express on its way to St. Louis, to whom he delivered his dispatches, and returned to camp, accompanied only by a Frenchman named Cabeneau, thus proving himself an efficient mountaineer at twenty years of age.

---

## 1831

As soon as the spring opened, sometime in March, the whole

2. Meek may have also gotten the story of his taking a dispatch to St. Louis misdated, since Fitzpatrick was leaving at about the same time for St. Louis. The name of the *Frenchman*—meaning French Canadian—Meek traveled back west with, *Cabeneau*, suggests that it may have been Jean Baptiste Charbonneau, son of Sacajawea. Harvey Tobie and Ann W. Hafen indicate, in the Hafen series, I, p. 317, and I, p. 213, respectively, that they think *Cabeneau* may have been that romantic fur trade figure. Charbonneau was traveling with the Bridger brigade the following year, 1832, which is as likely as Joe's date to be the right one. Until further evidence appears, the identity of *Cabeneau* remains speculation.

company started north again, for the Blackfoot country. But on the night of the third day out, they fell unawares into the neighborhood of a party of Crow Indians[3], whose spies discovered the company's horses feeding on the dry grass of a little bottom, and succeeded in driving off about three hundred head. Here was a dilemma to be in, in the heart of an enemy's country! To send the remaining horses after these, might be *sending the axe after the helve*, besides, most of them belonged to the free trappers, and could not be pressed into the service.

The only course remaining was to select the best men and dispatch them on foot, to overtake and retake the stolen horses. Accordingly, one hundred trappers were ordered on this expedition, among whom were Meek, Newell, and Antoine Godin, a half-breed and brave fellow, who was to lead the party.[4] Following the trail of the Crows for two hundred miles, traveling day and night, on the third day they came up with them on a branch of the Bighorn River. The trappers advanced cautiously, and being on the opposite side of the stream, on a wooded bluff, were enabled to approach close enough to look into their fort, and count the unsuspecting thieves. There were sixty of them, fine young braves, who believed that now they had made a start in life. Alas, for the vanity of human, and especially of Crow

---

3. Meek says, and said at the time, that the Crows stole three hundred head. Newell says 57 head.

4. Meek is vague about the route of the big brigade led by Bridger and Milton Sublette on this hunt of spring, 1831, leaving a big gap between the Yellowstone River and Pierre's Hole. Newell, who was also with the brigade, says Bridger led them elsewhere entirely—to the Laramie River and New Park—on the North Platte River—then to Snake River country at Bear River and Bear Lake, then to rendezvous in Cache Valley on Bear River. Though Meek says the rendezvous was on Green River, the consensus says otherwise.

expectations! Even then, while they were grouped around their fires, congratulating themselves on the sudden wealth which had descended upon them, as it were from the skies, an envious fate, in the shape of several roguish white trappers, was laughing at them and their hopes, from the overhanging bluff opposite them. And by and by, when they were wrapped in a satisfied slumber, two of these laughing rogues, Robert Newell, and Antoine Godin, stole under the very walls of their fort, and setting the horses free, drove them across the creek.

The Indians were awakened by the noise of the trampling horses, and sprang to arms. But Meek and his fellow trappers on the bluff fired into the fort with such effect that the Crows were appalled. Having delivered their first volley, they did not wait for the savages to recover from their recoil. Mounting in hot haste, the cavalcade of bare-back riders, and their drove of horses, were soon far away from the Crow fort, leaving the ambitious braves to finish their excursion on foot. It was afterward ascertained that the Crows lost seven men by that one volley of the trappers.

Flushed with success, the trappers yet found the backward journey more toilsome than the outward—for what with sleeplessness and fatigue, and bad traveling in melted snow, they were pretty well exhausted when they reached camp. Fearing, however, another raid from the thieving Crows, the camp got in motion again with as little delay as possible. They had not gone far, when Fitzpatrick turned back, with only one man, to go to St. Louis for supplies.

After the departure of Fitzpatrick, Bridger and Sublette completed their spring and summer campaign without any material loss in men or animals, and with considerable gain in beaver skins. Having once more visited the Yellowstone, they turned to the south again, crossing the mountains into Pierre's Hole, on to Snake River, thence to Salt River, thence to Bear River, and thence to Green River to rendezvous.

It was expected that Fitzpatrick would have arrived from St. Louis with the usual annual recruits and supplies of merchandise, in time for the summer rendezvous, but after waiting for some time in vain, Bridger and Sublette determined to send out a small party to look for him. The large number of men now employed, had exhausted the stock of goods on hand. The camp was without blankets and without ammunition, knives were not to be had, traps were scarce, but worse than all, the tobacco had given out, and alcohol was not! In such a case as this, what could a mountain man do?

To seek the missing Booshway became not only a duty, but a necessity, and not only a necessity of the physical man, but in an equal degree a need of the moral and spiritual man, which was rusting with the tedium of waiting. In the state of uncertainty in which the minds of the company were involved, it occurred to that of Frapp to consult a great *medicine man* of the Crows, one of those recruits filched from Mr. Ogden's party by whiskey the previous year.

Like all eminent professional men, the Crow chief required a generous fee, of the value of a horse or two, before he would begin to make *medicine*. This peculiar ceremony is pretty much alike among all the different tribes. It is observed first in the making of a medicine man, i.e. qualifying him for his profession, and afterward is practiced to enable him to heal the sick, to prophecy, and to dream dreams, or even to give victory to his people. To a medicine man was imputed great power, not only to cure, but to kill, and if, as it sometimes happened, the relatives of a sick man suspected the medicine man of having caused his death, by the exercise of evil powers, one of them, or all of them, pursued him to the death. Therefore, although it might be honorable, it was not always safe to be a great *medicine*.

The Indians placed a sort of religious value upon the practice of fasting, a somewhat curious fact, when it is remembered how many compulsory fasts they are obliged to endure, which must train them to think lightly of the deprivation of food. Those, however, who could endure voluntary abstinence long enough, were enabled to become very wise and very brave. The manner of making a *medicine* among some of the interior tribes, is in certain respects, similar to the practice gone through with by some preachers, in making a convert. A sort of camp-meeting is held, for several nights, generally about five, during which various dances are performed, with cries, and incantations, bodily exercises, singing, and nervous excitement, enough to make many patients, instead of one doctor. But the native's constitution is a strong one, and he holds out well. At last, however, one or more are overcome with the mysterious *power* which enters into them at that time, making, instead of a saint, only a superstitious Indian doctor.

The same sort of exercises which had made the Cree man a doctor were now resorted to, in order that he might obtain a more than natural sight, enabling him to see visions of the air, or at the least to endow him with prophetic dreams. After several nights of singing, dancing, hopping, screeching, beating of drums, and other more violent exercises and contortions, the exhausted medicine man fell off to sleep, and when he awoke he announced to Frapp that Fitzpatrick was not dead. He was on the road, some road, but not the right one etc. etc.

Thus encouraged, Frapp determined to take a party, and go in search of him. Accordingly, Meek, Reese, Ebarts, and Nelson, volunteered to accompany him. This party set out, first in the direction of Wind River, but not discovering any signs of the lost Booshway in that quarter, crossed over to the Sweetwater, and kept along down to the North Fork of the Platte, and thence to the Black Hills, where they found a beautiful country full of game, but not the hoped for train, with supplies. After waiting for a short time at the Black Hills, Frapp's party returned to the North Fork of the Platte, and were rejoiced to meet at last, the long-absent partner, with his pack train. Urged by Frapp, Fitzpatrick hastened forward, and came into camp on Powder River after winter had set in.

Fitzpatrick had a tale to tell the other partners, in explanation of his unexpected delay. When he had started for St. Louis in the month of March previous, he had hoped to have met the old partners, Capt. Sublette and Jedediah Smith, and to have obtained the necessary supplies from them, to furnish the Summer rendezvous with plenty. But these gentlemen, when he fell in with them, used certain arguments which induced him to turn back, and accompany them to Santa Fe, where they promised to furnish him goods, as he desired, and to procure for him an escort at that place. The journey had proven tedious, and unfortunate. They had several times been attacked by Indians, and Smith had been killed. While they were camped on a small tributary of the Simmaron River, Smith had gone a short distance from camp to procure water, and while at the stream was surprised by an ambush, and murdered on the spot, his murderers escaping unpunished. Sublette, now left alone in the business, finally furnished him, and he had at last made his way back to his Rocky Mountain camp.

But Fitzpatrick's content at being once more with his company was poisoned by the disagreeable proximity of a rival company. If he had annoyed Mr. Ogden of the Hudson's Bay Company, in the previous autumn, Major Vanderburg and Mr. Dripps, of the American Company, in their turn, annoyed him. This company had been on their heels, from the Platte River, and now were camped in the same neighborhood, using the Rocky Mountain Company as pilots to show them the country. As this was just what it was not for their interest to do, the Rocky Mountain Company raised camp, and fairly ran away from them, crossing the mountains to the Forks of the Snake River, where they wintered among the Nez Perces and Flathead Indians.

Sometime during this winter, Meek and Legarde, who had escaped from the Pawnees, made another expedition together, traveling three hundred miles on snowshoes, to the Bitter Root River, to look for a party of free trappers, whose beaver the company wished to secure. They were absent two months and a half, on this errand, and were entirely successful, passing a Blackfoot village in the night, but having no adventures worth recounting.[5]

5. Meek says he wintered with Fitzpatrick and Fraeb at the Forks of the Snake, among the Nez Percé and Flathead Indians. But the junction of Henry's and Lewis's Forks of the Snake is not in the country of those Indians. The junction of the Salmon and Lemhi Rivers, where Carson, Newell, and others indicate that this winter camp was, is in the right territory.

# 6

## 1832

n the following spring, the Rocky Mountain Fur Company commenced its march[1], first up Lewis's Fork, then on to Salt River, thence to Gray's River, and thence to Bear River. They fell in with the North American Fur Company on the latter river, with a large lot of goods, but no beaver. The American Company's resident partners were ignorant of the country, and were greatly at a loss where to look for the good trapping grounds. These gentlemen, Vanderburg and Dripps, were therefore inclined to keep an eye on the movements of the Rocky Mountain Company, whose leaders were acquainted with the whole region lying along the mountains, from the headwaters of the Colorado to the northern branches of the Missouri. On the other hand, the Rocky Mountain Company were anxious to *shake the*

1. Meek's account of the route of the spring hunt, led by Bridger, seems sound.

I *dust from off their feet*, which was trodden by the American Company, and to avoid the evils of competition in an Indian country. But they found the effort quite useless. The rival company had a habit of turning up in the most unexpected places, and taking advantage of the hard-earned experience of the Rocky Mountain Company's leaders. They tampered with the trappers, and ferreted out the secret of their next rendezvous. They followed on their trail, making them pilots to the trapping grounds, they sold goods to the Indians, and what was worse, to the hired trappers. In this way grew up that fierce conflict of interests, which made it *as much as his life was worth* for a trapper to suffer himself to be inveigled into the service of a rival company, which about this time or a little later, was at its highest, and which finally ruined the fur trade for the American companies in the Rocky Mountains.

Finding their rivals in possession of the ground, Bridger and Milton Sublette resolved to spend but a few days in that country. But so far as Sublette was concerned, circumstances ordered differently. A Rockway Chief, named Gray, and seven of his people, had accompanied the camp from Ogden's Hole, in the capacity of trappers. But during the sojourn on Bear River, there was a quarrel in camp on account of some indignity, real or fancied, which had been offered to the chief's daughter, and in the affray, Gray stabbed Sublette so severely that it was thought he must die.[2]

It thus fell out that Sublette had to be left behind, and Meek,

---

2. Meek tended Sublette in Ogden's Hole. For their route from there to Pierre's Hole to make sense, the incident with the Shoshone must have taken place on one of the westerly forks of the Green, perhaps Ham's Fork or Horse Creek.

who was his favorite, was left to take care of him while he lived, and bury him if he died, which trouble Sublette saved him, however, by getting well. But they had forty lonesome days to themselves after the camps had moved off—one on the heels of the other, to the great vexation of Bridger. Time passed slowly in Sublette's lodge, while waiting for his wound to heal. Day passed after day, so entirely like each other that the monotony alone seemed sufficient to invite death to an easy conquest. But the mountain man's blood, like the Indians, is strong and pure, and his flesh heals readily, therefore, since death would not have him, the wounded man was forced to accept of life in just this monotonous form. To him, Joe Meek was everything—hands, feet, physician, guard, caterer, hunter, cook, companion, friend. What long talks they had, when Sublette grew better: what stories they told, what little glimpses of a secret chamber in their hearts, and a better than the everyday spirit, in their bosoms, was revealed—as men will reveal such things in the isolation of sea-voyages, or the solitary presence of majestic nature.

To the veteran mountaineer there must have been something soothing in the care and friendship of the youth of twenty-two, with his daring disposition, his frankness, his cheerful humor, and his good looks—for our Joe was growing to be a maturely handsome man—tall, broad-shouldered, straight, with plenty of flesh, and none too much of it, a southerner's olive complexion, frank, dark eyes, and a classical nose and chin. What though in the matter of dress he was ignorant of the latest styles? Grace imparts elegance even to the trapper's beaver-skin cap and blanket capote.

At the end of forty days, as many as it took to drown a world, Sublette found himself well enough to ride, and the two set out on their search for camp. But now, other adventures awaited them. On a fork of Green River, they came suddenly upon a band of Snake Indians feeding their horses. As soon as the Snakes discovered the white men, they set up a yell, and made an instinctive rush for their horses. Now was the critical moment. One word passed between the travelers, and they made a dash past the savages, right into the village, and never slacked rein until they threw themselves from their horses at the door of the Medicine lodge. This is a large and fancifully decorated lodge, which stands in the centre of a village, and like the churches of Christians, is sacred. Once inside of this, the strangers were safe for the present, their blood could not be shed there.

The warriors of the village soon followed Sublette and Meek into their strange house of refuge. In half an hour, it was filled. Not a word was addressed to the strangers, nor by them to the Indians, who talked among themselves with a solemn eagerness, while they smoked the medicine pipe, as inspiration in their councils. Great was the excitement in the minds of the listeners, who understood the Snake tongue, as the question of their life or death was gravely discussed, yet in their countenances appeared only the utmost serenity. To show fear, is to whet an Indian's appetite for blood: coolness confounds and awes him when anything will.

If Sublette had longed for excitement, while an invalid in his lonely lodge on Bear River, he longed equally now for that blissful seclusion. Listening for, and hearing one's death-warrant from a band of blood-thirsty savages, could only prove with bitter sharpness how sweet was life, even the most uneventful. For hours, the council continued, and the majority favored the death sentence. But one old chief, called the good *Gotia*, argued long for an acquittal: he did not see the necessity of murdering two harmless travelers of the white race. Nothing availed, however, and just at sunset, their doom was fixed.

The only hope of escape was, that, favored by darkness, they might elude the vigilance of their jailers, and night, although so near, seemed ages away, even at sundown. Death being decreed, the warriors left the lodge one by one to attend to the preparation of the preliminary ceremonies. Gotia, the good, was the last to depart. As he left the Medicine lodge, he made signs to the captives to remain quiet until he should return, pointing upward to signify that there was a chance of life, and downward to show that possibly they must die.

What an age of anxiety was that hour of waiting! Not a word had been exchanged between the prisoners since the Indians entered the lodge, until now, and now very little was said, for speech would draw upon them the vigilance of their enemy, by whom they desired most ardently to be forgotten.

About dusk, there was a great noise, and confusion, and clouds of dust, in the south end of the village. Something was going wrong among the Indian horses. Immediately, all the village ran to the scene of the disorder, and at the same moment, Gotia, the good, appeared at the door of the Medicine lodge, beckoning the prisoners to follow him. With alacrity, they sprang up and after him, and were led across the stream, to a thicket on the opposite side, where their horses stood, ready to mount, in the charge of a young Indian girl. They did not stop for compliments, though had time been less precious, they might well have bestowed some moments of it in admiration of *Umentucken Tukutsey Undewatsey*, the Mountain Lamb. Soon after, the beautiful Snake girl became the wife of Milton Sublette, and after his return to the States, of the subject of this narrative, from which circumstance the incident above related, takes on something of the rosy hue of romance.

As each released captive received his bridle from the delicate hand of the Mountain Lamb, he sprang to the saddle. By this time, the chief had discovered that the strangers understood the Snake dialect. "Ride, if you wish to live," said he. "Ride without stopping, all night: and to-morrow linger not." With hurried thanks, our mountain men replied to this advice, and striking into a gallop, were soon far away from the Snake village. The next day at noon found them a hundred and fifty miles on their way to camp. Proceeding without further accident, they crossed the Teton Mountains, and joined the company at Pierre's Hole, after an absence of nearly four months.

Here they found the ubiquitous if not omnipresent American Fur Company encamped at the rendezvous of the Rocky Mountain Company. The partners, being anxious to be freed from this sort of espionage, and obstinate competition on their own ground, made a proposition to Vanderburg and Dripps to divide the country with them, each company to keep on its own territory. This proposition was refused by the American Company, perhaps because they feared having the poorer portion set off to themselves by their more experienced rivals. On this refusal, the Rocky Mountain Company determined to send an express to meet Capt. William Sublette, who was on his way out with a heavy stock of merchandise, and hurry him forward, lest the American Company should have the opportunity of disposing of its goods, when the usual gathering to rendezvous began. On this decision being formed, Fitzpatrick determined to go on this errand himself, which he accordingly did, falling in with Sublette, and Campbell, his associate, somewhere near the Black Hills. To them, he imparted his wishes and designs, and receiving the assurance of an early arrival at rendezvous, parted from them at the Sweetwater, and hastened back, alone, as he came, to prepare for business.

Captain Sublette hurried forward with his train, which consisted of sixty men with packhorses, three to a man. In company with him, was Mr. Nathaniel Wyeth, a history of whose fur-trading and salmon-fishing adventures has already been given. Captain Sublette had fallen in with Mr. Wyeth at Independence, Missouri, and finding him ignorant of the undertaking on which he was launched, offered to become pilot and traveling companion, an offer which was gratefully accepted.

The caravan had reached the foothills of the Wind River Mountains, when the raw recruits belonging to both these parties were treated to a slight foretaste of what Indian fighting would be, should they ever have to encounter it. Their camp was suddenly aroused at midnight by the simultaneous discharge of guns and arrows, and the frightful whoops and yells with which the savages make an attack. Nobody was wounded, however, but on springing to arms, the Indians fled, taking with them a few horses which their yells had frightened from their pickets. These marauders were Blackfeet, as Captain Sublette explained to Mr. Wyeth, their moccasin tracks having betrayed them, for as each tribe has a peculiar way of making or shaping the moccasin, the expert in Indian habits can detect the nationality of an Indian thief by his foot-print. After this episode of the night assault, the leaders redoubled their watchfulness, and reached their destination in Pierre's Hole about the first of July.

When Sublette arrived in camp, it was found that Fitzpatrick was missing. If the other partners had believed him to be with the captain, the captain expected to find him with them, but since neither could account to the other for his non-appearance, much anxiety was felt, and Sublette remembered with apprehension the visit he had received from Blackfeet. However, before anything had been determined upon with regard to him, he made his appearance in camp, in company with two Iroquois half-breeds, belonging to the camp, who had been out on a hunt.

Fitzpatrick had met with an adventure, as had been conjectured. While coming up the Green River valley, he described a small party of mounted men, whom he mistook for a company of trappers, and stopped to reconnoiter, but almost at the same moment, the supposed trappers, perceiving him, set up a yell that quickly undeceived him, and compelled him to flight. Abandoning his packhorse, he put the other to its topmost speed and succeeded in gaining the mountains, where in a deep and dark defile, he secreted himself until he judged the Indians had left that part of the valley. In this he was deceived, for no sooner did he emerge again into the open country, than he was once more pursued, and had to abandon his horse, to take refuge among the cliffs of the mountains. Here he remained for several days, without blankets or provisions, and with only one charge of ammunition, which was in his rifle, and kept for self-defense. At length, however, by frequent reconnoitering, he managed to elude his enemies, traveling by night, until he fortunately met with the two hunters from camp, and was conveyed by them to the rendezvous.[3]

All the parties were now safely in. The lonely mountain valley was populous with the different camps. The Rocky Mountain and American companies had their separate camps. Wyeth had his, a company of free trappers, fifteen in number, led by a man named Sinclair, from Arkansas, had the fourth, the Nez Perces and Flatheads. The allies of the Rocky Mountain Company, and the friends of the whites, had their lodges along all the streams, so that altogether there could not have been less than one thousand souls, and two or three thousand horses and mules gathered in this place.

"When the pie was opened then the birds began to sing." When Captain Sublette's goods were opened and distributed among the trappers and Indians, then began the usual gay carousal, and the *fast young men* of the mountains outvied each other in all manner of mad pranks. In the beginning of their spree, many feats of horsemanship and personal strength were exhibited, which were regarded with admiring wonder by the sober and inexperienced New Englanders under Mr. Wyeth's command. And as nothing stimulated the vanity of the mountain men like an audience of this sort, the feats they performed were apt to astonish themselves. In exhibitions of the kind, the free trappers took the lead, and usually carried off the palm, like the privileged class that they were.

But the horse-racing, fine riding, wrestling, and all the manlier sports, soon degenerated into the baser exhibitions of a

3. The fullest account of Fitzpatrick's adventure, allegedly in his own words, is in Leonard, p. 36-40.

*crazy drunk* condition. The vessel in which the trapper received and carried about his supply of alcohol was one of the small camp kettles. *Passing round* this clumsy goblet very freely, it was not long before a goodly number were in the condition just named, and ready for any mad freak whatever. It is reported by several of the mountain men that on the occasion of one of these *frolics*, one of their number seized a kettle of alcohol, and poured it over the head of a tall, lank, redheaded fellow, repeating as he did so the baptismal ceremony. No sooner had he concluded, than another man with a lighted stick, touched him with the blaze, when in an instant, he was enveloped in flames. Luckily, some of the company had sense enough left to perceive his danger, and began beating him with pack saddles to put out the blaze. But between the burning and the beating, the unhappy wretch nearly lost his life, and never recovered from the effects of his baptism by fire.

Beaver being plenty in camp, business was correspondingly lively, there being a great demand for goods. When this demand was supplied, as it was in the course of about three weeks, the different brigades were set in motion. One of the earliest to move was a small party under Milton Sublette, including his constant companion, Meek. With this company, no more than thirty in number, Sublette intended to explore the country to the southwest, then unknown to the fur companies, and to proceed as far as the Humboldt River in that direction.

On the 17th of July, they set out toward the south end of the valley, and having made but about eight miles the first day, camped that night near a pass in the mountains. Wyeth's party of raw New Englanders, and Sinclair's free trappers, had joined themselves to the company of Milton Sublette, and swelled the number in camp to about sixty men, many of them new to the business of mountain life.

Just as the men were raising camp for a start the next morning, a caravan was observed moving down the mountain pass into the valley. No alarm was at first felt, as an arrival was daily expected of one of the American Company's partisans, Mr. Fontenelle, and his company. But on reconnoitering with a glass, Sublette discovered them to be a large party of Blackfeet, consisting of a few mounted men, and many more men, women, and children, on foot. At the instant they were discovered, they set up the usual yell of defiance, and rushed down like a mountain torrent into the valley, flourishing their weapons, and fluttering their gay blankets and feathers in the wind. There was no doubt as to the warlike intentions of the Blackfeet in general, nor was it for a moment to be supposed that any peaceable overture on their part meant anything more than that they were not prepared to fight at that particular juncture, therefore, let not the reader judge too harshly of an act which under ordinary circumstances would have been infamous. In Indian fighting, every man is his own leader, and the bravest take the front rank. On this occasion, there were two of Sublette's men, one a half-breed Iroquois, the other a Flathead Indian, who had wrongs of their own to avenge, and they never let slip a chance of killing a Blackfoot. These two men rode forth alone to meet the enemy, as if to hold a *talk* with the principal chief, who advanced to meet them, bearing the pipe of peace. When the chief extended his hand, Antonio Godin, the half-breed, took it, but at the same moment he ordered the Flathead to fire, and the chief fell dead. The two trappers galloped back to camp, Antoine bearing for a trophy the scarlet blanket of his enemy.

This action made it impossible to postpone the battle, as the dead chief had meant to do by peaceful overtures, until the warriors of his nation came up. The Blackfeet immediately betook themselves to a swamp formed by an old beaver dam, and thickly overgrown with cottonwood and willow, matted together with tough vines. On the edge of this dismal covert the warriors skulked, and shot with their guns and arrows, while in its very midst the women employed themselves in digging a trench and throwing up a breastwork of logs, and whatever came to hand. Such a defense as the thicket afforded was one not easy to attack, with its unseen but certain dangers being sufficient to appall the stoutest heart.

Meantime, an express had been sent off to inform Captain Sublette of the battle, and summon assistance. Sinclair and his free trappers, with Milton Sublette's small company, were the only fighting men at hand. Mr. Wyeth, knowing the inefficiency of his men in an Indian fight, had them entrenched behind their packs, and there left them to take care of themselves, but charged them not to appear in open field. As for the fighting men, they stationed themselves in a ravine, where they could occasionally pick off a Blackfoot, and waited for reinforcements.

Great was the astonishment of the Blackfeet, who believed they had only Milton Sublette's camp to fight, when they beheld first one party of white men and then another, and not only whites, but Nez Percé and Flatheads came galloping up the valley. If before it had been a battle to destroy the whites, it was now a battle to defend themselves. Previous to the arrival of Captain Sublette, the opposing forces had kept up only a scattering fire, in which nobody on the side of the trappers had been either killed or wounded. But when the impetuous captain arrived on the battlefield, he prepared for less guarded warfare. Stripped as if for the prize-ring, and armed *cap-a-pie*, he hastened to the scene of action, accompanied by his intimate friend and associate in business, Robert Campbell.

At sight of the reinforcements, and their vigorous movements, the Indians at the edge of the swamp fell back within their fort. To dislodge them was a dangerous undertaking, but Captain Sublette was determined to make the effort. Finding the trappers generally disinclined to enter the thicket, he set the example, together with Campbell, and thus induced some of the free trappers, with their leader, Sinclair, to emulate his action. However, the others took courage at this, and advanced near the swamp, firing at random at their invisible foe, who, having the advantage of being able to see them, inflicted some wounds on the party.

The few white *braves* who had resolved to enter the swamp, made their wills as they went, feeling that they were upon perilous business. Sublette, Campbell, and Sinclair succeeded in penetrating the thicket without alarming the enemy, and came at length to a more open space from whence they could get a view of the fort. From this they learned that the women and children had retired to the mountains, and that the fort was a slight affair, covered with buffalo robes and blankets to keep out prying eyes. Moving slowly on, some slight accident betrayed their vicinity, and the next moment, a shot struck Sinclair, wounding him mortally. He spoke to Campbell, requesting to be taken to his brother. By this time, some of the men had come up, and he was given in charge to be taken back to camp. Sublette then pressed forward, and seeing an Indian looking through an aperture, aimed at him with fatal effect. No sooner had he done so, and pointed out the opening to Campbell, than he was struck with a ball in the shoulder, which nearly prostrated him, and turned him so faint that Campbell took him in his arms and carried him, assisted by Meek, out of the swamp. At the same time one of the men received a wound in the head. The battle was now carried on with spirit, although from the difficulty of approaching the fort, the firing was very irregular.

The mountaineers who followed Sublette, took up their station in the woods on one side of the fort, and the Nez Perces, under Wyeth, on the opposite side, which accidental arrangement, though it was fatal to many of the Blackfeet in the fort, was also the occasion of loss to themselves by the cross-fire. The whites, being constantly reinforced by fresh arrivals from the rendezvous, were soon able to silence the guns of the enemy, but they were not able to drive them from their fort, where they remained silent and sullen after their ammunition was exhausted.

Seeing that the women of the Nez Perces and Flatheads were gathering up sticks to set fire to their breastwork of logs, an old chief proclaimed in a loud voice from within, the startling intelligence that there were four hundred lodges of his people close at hand, who would soon be there to avenge their deaths, should the whites choose to reduce them to ashes. This harangue, delivered in the usual high-flown style of Indian oratory, either was not clearly understood, or was wrongly interpreted, and the impression got abroad that an attack was being made on the great encampment. This intelligence occasioned a diversion, and a division of forces, and for while a small party was left to watch the fort, the rest galloped in hot haste to the rescue of the main camp. When they arrived, they found it had been a false alarm, but it was too late to return that night, and the several camps remained where they were until the next day.

Meantime, the trappers left to guard the fort remained stationed within the wood all night, firmly believing they had their enemy *corralled*, as the horsemen of the plains would say. On the return, in the morning, of their comrades from the main camp, they advanced cautiously up to the breastwork of logs, and behold! Not a buffalo skin or red blanket was to be seen! Through the crevices among the logs was seen an empty fort. On making this discovery, there was much chagrin among the white trappers, and much lamentation among the Indian allies, who had abandoned the burning of the fort expressly to save for themselves the fine blankets and other goods of their hereditary foes.

From the reluctance displayed by the trappers, in the beginning of the battle, to engage with the Indians while under cover of the woods, it must not be inferred that they were lacking in courage. They were too well informed in Indian modes of warfare to venture recklessly into the den of death, which a savage ambush was quite sure to be. The very result which attended the impetuosity of their leaders, in the death of Sinclair and the wounding of Captain Sublette, proved them not over cautious.

On entering the fort, the dead bodies of ten Blackfeet were found, besides others dead outside the fort, and over thirty horses, some of which were recognized as those stolen from Sublette's night camp on the other side of the mountains, besides those abandoned by Fitzpatrick. Doubtless, the rascals had followed his trail to Pierre's Hole, not thinking, however, to come upon so large a camp as they found at last. The savage garrison which had so cunningly contrived to elude the guard set upon them, carried off some of their wounded, and perhaps, also some of their dead, for they acknowledged afterward a much larger loss than appeared at the time. Besides Sinclair, there were five other white men killed, one-half-breed, and seven Nez Percé . About the same number of whites and their Indian allies were wounded.

An instance of female devotion is recorded by Bonneville's historian as having occurred at this battle. On the morning following it, as the whites were exploring the thickets about the fort, they discovered a Blackfoot woman leaning silent and motionless against a tree. According to Mr. Irving, whose fine feeling for the sex would incline him to put faith in this bit of romance, "Their surprise at her lingering here alone, to fall into the hands of her enemies, was dispelled when they saw the corpse of a warrior at her feet. Either she was so lost in grief as not to perceive their approach, or a proud spirit kept her silent and motionless. The Indians set up a yell on discovering her, and before the trappers could interfere, her mangled body fell upon the corpse which she had refused to abandon." This version is true in the main incidents, but untrue in the sentiment. The woman's leg had been broken by a ball, and she was unable to move from the spot where she leaned. When the trappers approached her, she stretched out her hands supplicatingly, crying out in a wailing voice, "Kill me! Kill me! O white men, kill me!" But this the trappers had no disposition to do. While she was entreating them, and they refusing, a ball from some vengeful Nez Percé or Flathead put an end to her sufferings.

Still remembering the threats of the Blackfoot chief, that four hundred lodges of his brethren were advancing on the valley, all the companies returned to rendezvous, and remained for several days, to see whether an attack should take place. But if there had ever been any such intention on the part of the Blackfoot nation, the timely lesson bestowed on their advance guard had warned them to quit the neighborhood of the whites.

Captain Sublette's wound was dressed by Mr. Wyeth's physician, and although it hindered his departure for St. Louis for some time, it did not prevent his making his usual journey later in the season. It was as well, perhaps, that he did not set out earlier, for of a party of seven who started for St. Louis a few days after the battle, three were killed in Jackson's Hole, where they fell in with the four hundred warriors with whom the Blackfoot chief threatened the whites at the battle of Pierre's Hole.[4] From the story of the four survivors who escaped and returned to camp, there could no longer be any doubt that the big village of the Blackfeet had actually been upon the trail of Capt. Sublette, expecting an easy victory when they should overtake him. How they were disappointed by the reception met with by the advance camp, has already been related.

---

4. A good account of the Battle of Pierre's Hole is in Irving, p. 58-63. The editor is unable to discover the meaning of *armed cap-a-pie*.

# 7

## 1832

On the 23rd of July, Milton Sublette's brigade and the company of Mr. Wyeth again set out for the southwest, and met no more serious interruptions while they traveled in company.[1] On the headwaters of the Humboldt River they separated, Wyeth proceeding north to the Columbia, and Sublette continuing on into a country hitherto untraversed by American trappers.[2]

It was the custom of a camp on the move to depend chiefly on the men employed as hunters to supply them with game, the sole support of the mountaineers. When this failed, the stock on hand was soon exhausted, and the men reduced to famine. This

---

1. Fall hunt of 1832 led by Milton Sublette: The route seems to have been just as Meek describes it here.

2. Probably Wyeth separated from Sublette's outfit near the mouth of the Owyhee River at the Snake.

was what happened to Sublette's company in the country where they now found themselves, between the Owyhee and Humboldt Rivers. Owing to the arid and barren nature of these plains, the largest game to be found was the beaver, whose flesh proved to be poisonous, from the creature having eaten of the wild parsnip in the absence of its favorite food. The men were made ill by eating of beaver flesh, and the horses were greatly reduced from the scarcity of grass and the entire absence of the cottonwood.

In this plight Sublette found himself, and finally resolved to turn north, in the hope of coming upon some better and more hospitable country. The sufferings of the men now became terrible, both from hunger and thirst. In the effort to appease the former, everything was eaten that could be eaten, and many things at which the well-fed man would sicken with disgust. "I have," says Joe Meek, "held my hands in an ant-hill until they were covered with the ants, then greedily licked them off. I have taken the soles off my moccasins, crisped them in the fire, and eaten them. In our extremity, the large black crickets which are found in this country were considered game. We used to take a kettle of hot water, catch the crickets and throw them in, and when they stopped kicking, eat them. That was not what we called *cant tickup ko hanch*—meaning good meat, my friend—but it kept us alive."

Equally abhorrent expedients were resorted to in order to quench thirst, some of which would not bear mention. In this condition, and exposed to the burning suns and the dry air of the desert, the men now so nearly exhausted, began to prey upon their almost equally exhausted animals. At night when they made their camp, by mutual consent a mule was bled, and a soup made from its blood. About a pint was usually taken, when two or three would mess together upon this reviving, but scanty and not very palatable dish. But this mode of subsistence could not be long depended on, as the poor mules could ill afford to lose blood in their famishing state, nor could the men afford to lose their mules where there was a chance of life: therefore hungry as they were, the men were cautious in this matter, and it generally caused a quarrel when a man's mule was selected for bleeding by the others.

A few times a mule had been sacrificed to obtain meat, and in this case the poorest one was always selected, so as to economise the chances for life for the whole band. In this extremity, after four days of almost total abstinence and several weeks of famine, the company reached the Snake River, about fifty miles above the fishing falls, where it boils and dashes over the rocks, forming very strong rapids. Here, the company camped, rejoiced at the sight of the pure mountain water, but still in want of food. During the march a horse's back had become sore from some cause—probably, his rider thought, because the saddle did not set well—and although that particular animal was selected to be sacrificed on the morrow, as one that could best be spared, he set about taking the stuffing out of his saddle and re-arranging the padding. While engaged in this considerate labor, he uttered a cry of delight and held up to view a large brass pin, which had accidentally got into the stuffing, when the saddle was made, and had been the cause of all the mischief to his horse.

The same thought struck all who saw the pin: it was soon converted into a fish-hook, a line was spun from horsehair, and in a short time there were trout enough caught to furnish them a hearty and a most delicious repast. "In the morning," says Meek, "we went on our way rejoicing," each man with the *five fishes* tied to his saddle, if without any *loaves*. This was the end of their severest suffering, as they had now reached a country where absolute starvation was not the normal condition of the inhabitants, and which was growing more and more bountiful, as they neared the Rocky Mountains, where they at length joined camp, not having made a very profitable expedition.

It may seem incredible to the reader that any country so poor as that in which our trappers starved could have native inhabitants. Yet such was the fact, and the people who lived in and who still inhabit this barren waste, were called *Diggers*, from their mode of obtaining their food—a few edible roots growing in low grounds, or marshy places. When these fail them, they subsist as did our trappers, by hunting crickets and field mice.

Nothing can be more abject than the appearance of the Digger Indian, in the fall, as he roams about, without food and without weapons, save perhaps a bow and arrows, with his eyes fixed upon the ground, looking for crickets! So despicable is he, that he has neither enemies nor friends, and the neighboring tribes do not condescend to notice his existence, unless indeed he should come in their way, when they would not think it more than a mirthful act to put an end to his miserable existence. And so it must be confessed the trappers regarded him. When Sublette's party first struck the Humboldt[3], Wyeth's being still with them, Joe Meek one day shot a Digger who was prowling about a stream where his traps were set.

---

3. Wyeth was doubtless not with Meek and Sublette when they reached the Humboldt, the New England trader having followed the Snake north to the Columbia.

"Why did you shoot him?" asked Wyeth.

"To keep him from stealing traps."

"Had he stolen any?"

*"No, but he looked as if he was going to!"*

This recklessness of life very properly distressed the just-minded New Englander. Yet it was hard for the trappers to draw lines of distinction so nice as his. If a tribe was not known to be friendly, it was a rule of necessity to consider it unfriendly. The abjectness and cowardice of the Diggers was the fruit of their own helpless condition. That they had the savage instinct, held in check only by circumstances, was demonstrated about the same time that Meek shot one, by his being pursued by four of them when out trapping alone, and only escaping at last by the assistance of one of his comrades who came to the rescue. They could not fight, like the Crows and Blackfeet, but they could steal and murder, when they had a safe opportunity.

It would be an interesting study, no doubt, to the philanthropist, to ascertain in how great a degree the habits, manners, and morals of a people are governed by their resources, especially by the quality and quantity of their diet. But when diet and climate are both taken into consideration, the result is striking.

The character of the Blackfeet who inhabited the good hunting grounds on the eastern side of the Rocky Mountains, is already pretty well given. They were tall, sinewy, well-made fellows, good horsemen, and good fighters, though inclined to marauding and murdering. They dressed comfortably and even handsomely, as dress goes among savages, and altogether were more to be feared than despised.

The Crows resembled the Blackfeet, whose enemies they were, in all the before-mentioned traits, but were, if possible, even more predatory in their habits. Unlike the Blackfeet, however, they were not the enemies of *all* mankind, and even were disposed to cultivate some friendliness with the white traders and trappers, in order, as they acknowledged, to strengthen their own hands against the Blackfeet. They, too, inhabited a good country, full of game, and had horses in abundance. These were the mountain tribes.

Comparing these with the coast tribes, there was a striking difference. The natives of the Columbia were not a tall and robust people, like those east of the Rocky Mountains, who lived by hunting. Their height rarely exceeded five feet six inches, their forms were good, rather inclining to fatness, their faces round, features coarse, but complexion light, and their eyes large and intelligent. The custom of flattening their heads in infancy gave them a grotesque and unnatural appearance, otherwise they could not be called ill-looking. On the first advent of white men among them, they were accustomed to go entirely naked, except in winter, when a panther skin, or a mantle of other skins sewed together, served to protect them from the cold: or if the weather was rainy, as it generally was in that milder climate, a long mantle of rush mats, like the toga of the ancient Romans, took the place of that made of skins. To this was added a conical hat, woven of fibrous roots, and gaily painted.

For defensive armor, they were provided with a tunic of elkskin double, descending to the ankles, with holes in it for the arms, and quite impenetrable to arrows. A helmet of similar material covered the head, rendering them like Achilles, invulnerable except in the heels. In this secure dress, they went to battle in their canoes, notice being first given to the enemy of the intended attack. Their battles might, therefore, be termed compound duels, in which each party observed great punctiliousness and decorum. Painted and armor-encased, the warriors in two flotillas of canoes were rowed to the battleground by their women, when the battle raged furiously for some time, not, however, doing any great harm to either side. If anyone chanced to be killed, that side considered itself beaten, and retired from the conflict to mourn over and bury the estimable and departed brave. If the case was a stubborn one, requiring several days fighting, the opponents encamped near each other, keeping up a confusion of cries, taunts, menaces, and raillery, during the whole night, after which they resumed the conflict, and continued it until one was beaten. If a village was to be attacked, notice being received, the women and children were removed, and if the village was beaten, they made presents to their conquerors. Such were the decorous habits of the warriors of the Lower Columbia.

These were the people who lived almost exclusively by fishing, and whose climate was a mild and moist one. Fishing, in which both sexes engaged about equally, was an important accomplishment, since it was by fish they lived in this world, and by being good fishermen that they had hopes of the next one. The houses in which they lived, instead of being lodges made of buffalo skins, were of a large size and very well constructed, being made out of cedar planks. An excavation was first made in the earth two or three feet deep, probably to secure greater warmth in winter. A double row of cedar posts was then planted firmly all round the excavation, and between these the planks were laid, or, sometimes cedar bark, so overlapped as to exclude the rain and wind. The ridge-pole of the roof was supported on a row of taller posts, passing through the centre of the building, and notched to receive it. The rafters were then covered with planks or bark, fastened down with ropes made of the fiber of the cedar bark. A house made in this manner, and often a hundred feet long by thirty or forty wide, accommodated several families, who each had their separate entrance and fire-place, with the entrance being by a low oval-shaped door, and a flight of steps.

The canoes of these people were each cut out of a single log of cedar, and were often thirty feet long and five wide at midships. They were gaily painted, and their shape was handsome, with a very long bow so constructed as to cut the surf in landing with the greatest ease, or the more readily to go through a rough sea. The oars were about five feet long, and bent in the shape of a crescent, which shape enabled them to draw them edgewise through the water with little or no noise—this noiselessness being an important quality in hunting the sea otter, which is always caught sleeping on the rocks.

The single instrument which sufficed to build canoes and houses was the chisel, generally being a piece of old iron obtained from some vessel and fixed in a wooden handle. A stone mallet aided them in using the chisel, and with this simple *kit* of tools they contrived to manufacture plates, bowls, carved oars, and many ornamental things.

Like the men of all savage nations, they made slaves of their captives, and their women. The dress of the latter consisted merely of a short petticoat, manufactured from the fiber of the cedar bark, previously soaked and prepared. This material was worked into a fringe, attached to a girdle, and only long enough to reach the middle of the thigh. When the season required it, they added a mantle of skins. Their bodies were anointed with fish-oil, and sometimes painted with red ochre in imitation of the men. For ornaments, they wore strings of glass beads, and also of a white shell found on the northern coast, called *haiqua*. Such were the *Chinooks*, who lived upon the coast.

Farther up the river, on the eastern side of the Cascade range of mountains, a people lived, the same, yet different from the Chinooks. They resembled them in form, features, and manner of getting a living. But they were more warlike and more enterprising. They even had some notions of commerce, being traders between the coast Indians and those to the east of them. They too were great fishermen, but used the net instead of fishing in boats. Great scaffoldings were erected every year at the narrows of the Columbia, known as the Dalles, where, as the salmon passed up the river in the spring, in incredible numbers, they were caught and dried. After drying, the fish were then pounded fine between two stones, pressed tightly into packages or bales of about a hundred pounds, covered with matting, and corded up for transportation. The bales were then placed in storehouses built to receive them, where they awaited customers.

By and by, there came from the coast other Indians, with different varieties of fish, to exchange for the salmon in the Wish-ram warehouses. And by and by there came from the plains to the eastward, others who had horses, camas-root, bear-grass, fur robes, and whatever constituted the wealth of the mountains and plains, to exchange for the rich and nutritious salmon of the Columbia. These Wish-ram Indians were sharp traders, and usually made something by their exchanges, so that they grew rich and insolent, and it was dangerous for the unwary stranger to pass their way. Of all the tribes of the Columbia, they perpetrated the most outrages upon their neighbors, the passing traveler, and the stranger within their gates.

Still farther to the east, on the great grassy plains, watered by beautiful streams, coming down from the mountains, lived the Cayuses, Yakimas, Nez Perces, Wallah-Wallahs, and Flatheads, as different in their appearance and habits as their different modes of living would naturally make them. Instead of having many canoes, they had many horses, and in place of drawing the fishing net, or trolling lazily along with hook and line, or spearing fish from a canoe, they rode pell-mell to the chase, or sallied out to battle with the hostile Blackfeet, whose country lay between them and the good hunting grounds, where the great herds of buffalo were. Being Nimrods by nature, they were dressed in complete suits of skins, instead of going naked, like their brethren in the lower country. Being wandering and pastoral in their habits, they lived in lodges, which could be planted every night and raised every morning.

Their women, too, were good riders, and comfortably clad in dressed skins, kept white with chalk. So wealthy were some of the chiefs that they could count their fifteen hundred head of horses grazing on their grassy uplands. Horse-racing was their delight, and betting on them their besetting vice. For bridles they used horsehair cords, attached around the animal's mouth. This was sufficient to check him, and by laying a hand on this side or that of the horse's neck, the rider could wheel him in either direction. The simple and easy-fitting saddle was a stuffed deer-skin, with stirrups of wood, resembling in shape those used by the Mexicans, and covered with deerskin sewed on wet, so as to tighten in drying. The saddles of the women were furnished with a pair of deer's antlers for the pommel.

In many things, their customs and accoutrements resembled those of the Mexicans, from whom, no doubt, they were borrowed. Like the Mexican, they threw the lasso to catch the wild horse. Their horses, too, were of Mexican stock, and many of them bore the brand of that country, having been obtained in some of their not-infrequent journeys into California and New Mexico.

As all the wild horses of America are said to have sprung from a small band, turned loose upon the plains by Cortez, it would be interesting to know at what time they came to be used by the northern Indians, or whether the horse and the Indian did not emigrate together. If the horse came to the Indian, great must have been the change effected by the advent of this new element in the savage's life. It is impossible to conceive, however, that the Indian ever could have lived on these immense plains, barren of everything but wild grass, without his horse. With him he does well enough, for he not only *lives on horseback*, by which means he can quickly reach a country abounding in game, but he literally lives on horseflesh, when other game is scarce.

Curious as the fact may seem, the Indians at the mouth of the Columbia and those of New Mexico speak languages similar in construction to that of the Aztecs, and from this fact, and the others before mentioned, it may be very fairly inferred that difference of circumstances and localities have made of the different tribes what they are.

As to the Indian's moral nature, that is pretty much alike everywhere, and with some rare exceptions, the rarest of which is, perhaps, the Flathead and Nez Perces nations, all are cruel, thieving, and treacherous. The Indian gospel is literally the *gospel of blood*, an *eye for an eye, and a tooth for a tooth.*

Vengeance is as much a commandment to him as any part of the Decalogue is to the Christian. But we have digressed far from our narrative, and as it will be necessary to refer to the subject of the moral code of savages further on in our narrative, we leave it for the present.

After the incident of the pin and the fishes, Sublette's party kept on to the north, coursing along up Payette's River to Payette Lake, where he camped, and the men went out trapping. A party of four, consisting of Meek, Antoine Godin, Louis Leaugar, and Small, proceeded to the north as far as the Salmon River and beyond, to the head of one of its tributaries, where the present city of Florence is located. While camped in this region, three of the men went out one day to look for their horses, which had strayed away, or been stolen by the Indians. During their absence, Meek, who remained in camp, had killed a fine fat deer, and was cooking a portion of it, when he saw a band of about a hundred Indians approaching, and so near were they that flight was almost certainly useless. But, as a hundred against one was very great odds, and running away from them would not increase their number, while it gave him something to do in his own defense, he took to his heels and ran as only a mountain man can run. Instead, however, of pursuing him, the practical-minded braves set about finishing his cooking for him, and soon had the whole deer roasting before the fire.

This procedure provoked the gastronomic ire of our trapper, and after watching them for some time from his hiding place, he determined to return and share the feast. On reaching camp again, and introducing himself to his not over-scrupulous visitors, he found they were from the Nez Perces tribe inhabiting that region, who, having been so rude as to devour his stock of provisions, invited him to accompany them to their village, not a great way off, where they would make some return for his involuntary hospitality. This he did, and there found his three comrades and all their horses. While still visiting at the Nez Perces village, they were joined by the remaining portion of Sublette's command, when the whole company started south again. Passing Payette's lake to the east, traversing the Boise Basin, going to the headwaters of that river, thence to the Malade, thence to Godin's River, and finally to the Forks of the Salmon, where they found the main camp. Captain Bonneville, of whose three years wanderings in the wilderness Mr. Irving has given a full and interesting account, was encamped in the same neighborhood, and had built there a small fort or trading-house, and finally wintered in the neighborhood.

An exchange of men now took place, and Meek went east of the mountains under Fitzpatrick and Bridger. When these famous leaders had first set out for the summer hunt, after the battle of Pierre's Hole, their course had been to the headwaters of the Missouri, to the Yellowstone Lake, and the forks of the Missouri, some of the best beaver grounds known to them. But finding their steps dogged by the American Fur Company, and not wishing to be made use of as pilots by their rivals, they had flitted about for a time like an Arab camp, in the endeavor to blind them, and finally returned to the west side of the mountains, where Meek fell in with them.

Exasperated by the perseverance of the American Company, they had come to the determination of leading them a march which should tire them of the practice of keeping at their heels. They therefore planned an expedition, from which they expected no other profit than that of shaking off their rivals. Taking no pains to conceal their expedition, they rather held out the bait to the American Company, who, unsuspicious of their purpose, took it readily enough. They led them along across the mountains, and on to the headwaters of the Missouri. Here, packing up their traps, they tarried not for beaver, nor even tried to avoid the Blackfeet, but pushed right ahead, into the very heart of their country, keeping away from any part of it where beaver might be found, and going away on beyond, to the elevated plains, quite destitute of that small but desirable game, but followed through it by their rivals.

However justifiable on the part of trade this movement of the Rocky Mountain Company might have been, it was a cruel device as it concerned the inexperienced leaders of the other company, one of whom lost his life in consequence. Not knowing of their danger, they only discovered their situation in the midst of Blackfeet, after discovering the ruse that had been played upon them. They then halted, and being determined to find beaver, divided their forces and set out in opposite directions for that purpose. Unhappily, Major Vanderburg took the worst possible direction for a small party to take, and had not traveled far when his scouts came upon the still smoking campfires of a band of Indians who were returning from a buffalo hunt. From the *signs* left behind them, the scout judged that they had become aware of the near neighborhood of white men, and from their having stolen off, he judged that they were only gone for others of their nation, or to prepare for war.

But Vanderburg, with the fool-hardiness of one not *up to Blackfeet*, determined to ascertain for himself what there was to fear, and taking with him half a score of his followers, put himself upon their trail, galloping hard after them, until, in his rashness, he found himself being led through a dark and deep defile, rendered darker and gloomier by overhanging trees. In the midst of this dismal place, just where an ambush might have been expected, he was attacked by a horde of savages, who rushed upon his little party with whoops and frantic gestures, intended not only to appall the riders, but to frighten their horses, and thus make surer their bloody butchery. It was but the work of a few minutes to consummate their demoniac purpose. Vanderburg's horse was shot down at once, falling on his rider, whom the Indians quickly dispatched. One or two of the men were instantly tomahawked, and the others wounded while making their escape to camp. The remainder of Vanderburg's company, on learning the fate of their leader, whose place there was no one to fill, immediately raised camp and fled with all haste to the encampment of the Pends Oreille Indians for assistance. Here they waited, while those Indians, a friendly tribe, made an effort to recover the body of their unfortunate leader, but the remains were never recovered, probably having first been fiendishly mutilated, and then left to the wolves.[4]

Fitzpatrick and Bridger, finding they were no longer pursued by their rivals, as the season advanced, began to retrace their

4. The incidents of Drips and Vanderburgh following the brigade led by Fitzpatrick and Bridger, through Vanderburgh's death, is told from the point of view of a Vanderburgh man in Ferris, p. 159-178. The route of the combined parties was, according to Harvey Carter in the Hafen series, VII, p. 319, from Pierre's Hole "to Deer Lodge Valley, the Hellgate, or Clark's Fork of the Columbia, the Big Blackfoot, across the Continental Divide to the Great Falls of the Missouri and thence to the Three Forks of the Missouri." Then Vanderburgh separated his party from Drips, Bridger, and Fitzpatrick and went to trap the Ruby River. Vanderburgh and company were ambushed as they descended from Ennis Pass toward Alder Gulch.

steps toward the good trapping grounds. Being used to Indian wiles and Blackfeet marauding and ambushes, they traveled in close columns, and never camped or turned out their horses to feed, without the greatest caution. Morning and evening scouts were sent out to beat up every thicket or ravine that seemed to offer concealment to a foe, and the horizon was searched in every direction for signs of an Indian attack. The complete safety of the camp being settled almost beyond a peradventure, the horses were turned loose, though never left unguarded.

It was not likely, however, that the camp should pass through the Blackfoot country without any encounters with that nation. When it had reached the headwaters of the Missouri, on the return march, a party of trappers, including Meek, discovered a small band of Indians in a bend of the lake, and thinking the opportunity for sport a good one, commenced firing on them. The Indians, who were without guns, took to the lake for refuge, while the trappers entertained themselves with the rare amusement of keeping them in the water, by shooting at them occasionally. But it chanced that these were only a few stragglers from the main Blackfoot camp, which soon came up and put an end to the sport by putting the trappers to flight in their turn. The trappers fled to camp, the Indians pursuing, until the latter discovered that they had been led almost into the large camp of the whites. This occasioned a halt, the Blackfeet not caring to engage with superior numbers.

In the pause which ensued, one of the chiefs came out into the open space, bearing the peace pipe, and Bridger also advanced to meet him, but carrying his gun across the pommel of his saddle. He was accompanied by a young Blackfoot woman, wife of a Mexican in his service, as interpreter. The chief extended his hand in token of amity, but at that moment, Bridger saw a movement of the chiefs, which he took to mean treachery, and cocked his rifle. But the lock had no sooner clicked than the chief, a large and powerful man, seized the gun and turned the muzzle downward, when the contents were discharged into the earth. With another dexterous movement, he wrested it from Bridger's hand, and struck him with it, felling him to the ground. In an instant, all was confusion. The noise of whoops, yells, of firearms, and of running hither and thither, gathered like a tempest. At the first burst of this demoniac blast, the horse of the interpreter became frightened, and by a sudden movement, unhorsed her, wheeling and running back to camp. In the melee which now ensued, the woman was carried off by the Blackfeet, and Bridger was wounded twice in the back with arrows. A chance-medley fight now ensued, continuing until night put a period to the contest. So well matched were the opposing forces, that each fought with caution, firing from the cover of thickets and from behind rocks, neither side doing much execution. The loss on the part of the Blackfeet was nine warriors, and on that of the whites, three men and six horses.

As for the young Blackfoot woman, whose people retained her a prisoner, her lamentations and struggles to escape and return to her husband and child so wrought upon the young Mexican, who was the pained witness of her grief, that he took the babe in his arms, and galloped with it into the heart of the Blackfoot camp, to place it in the arms of the distracted mother. This daring act, which all who witnessed believed would cause his death, so excited the admiration of the Blackfoot chief, that he gave him permission to return, unharmed, to his own camp. Encouraged by this clemency, Loretta begged to have his wife restored to him, relating how he had rescued her, a prisoner, from the Crows, who would certainly have tortured her to death. The wife added her entreaties to his, but the chief sternly bade him depart, and as sternly reminded the Blackfoot girl that she belonged to his tribe, and could not go with his enemies. Loretta was therefore compelled to abandon his wife and child, and return to camp.

It is, however, gratifying to know that so true an instance of affection in savage life was finally rewarded, and that when the two rival fur companies united, as they did in the following year, Loretta was permitted to go to the American Company's fort on the Missouri, in the Blackfoot country, where he was employed as interpreter, assisted by his Blackfoot wife.

Such were some of the incidents that signalized this campaign in the wilderness, where two equally persistent rivals were trying to outwit one another. Subsequently, when several years of rivalry had somewhat exhausted both, the Rocky Mountain and American companies consolidated, using all their strategy thereafter against the Hudson's Bay Company, and any new rival that chanced to enter their hunting grounds.

After the fight above described, the Blackfeet drew off in the night, showing no disposition to try their skill the next day against such experienced Indian fighters as Bridger's brigade had shown themselves. The company continued in the Missouri country, trapping and taking many beaver, until it reached the Beaver Head Valley, on the headwaters of the Jefferson fork of the Missouri. Here, the lateness of the season compelled a return to winter-quarters, and by Christmas all the wanderers were gathered into camp at the Forks of the Snake River.

---

## 1833

In the latter part of January, it became necessary to move to the junction of the Portneuf to subsist the animals. The main body of the camp had gone on in advance, while some few, with pack-horses, or women with children, were scattered along the trail. Meek, with five others, had been left behind to gather up some horses that had strayed. When about a half day's journey from camp, he overtook *Umentucken*, the Mountain Lamb, now the wife of Milton Sublette, with her child, on horseback. The weather was terribly cold, and seeming to grow colder. The naked plains afforded no shelter from the piercing winds, and the air fairly glittered with frost. Poor Umentucken was freezing, but more troubled about her babe than herself. The camp was far ahead, with all the extra blankets, and the prospect was imminent that they would perish. Our gallant trapper had thought himself very cold until this moment, but what were his sufferings compared to those of the Mountain Lamb and her little Lambkin? Without an instant's hesitation, he divested himself of his blanket capote, which he wrapped round the mother and child, and urged her to hasten to camp. For himself, he could not hasten, as he had the horses in charge, but all that fearful afternoon rode naked above the waist, exposed to the wind, and the fine, dry, icy hail, which filled the air as with diamond needles, to pierce the skin, and probably, to the fact that the hail *was* so stinging, was owing the fact that his blood did not congeal.

"O what a day was that!" said Meek to the writer. "Why, the air war thick with fine, sharp hail, and the sun shining, too! not one sun only, but three suns—there were *three* suns! And when night came on, the northern lights blazed up the sky! It was the most beautiful sight I ever saw. That is the country for northern lights!"

When some surprise was expressed that he should have been obliged to expose his naked skin to the weather, in order to save Umentucken— "In the mountains," he answered, "we do not have many garments. Buckskin breeches, a blanket capote, and a beaver-skin cap makes up our rig."

"You do not need a laundress, then? But with such clothing, how could you keep free of vermin?"

"We didn't always do that. Do you want to know how we got rid of lice in the mountains? We just took off our clothes and laid them on an ant-hill, and you ought to see how the ants would carry off the lice!"

But to return to our hero, frozen, or nearly so. When he reached camp at night, so desperate was his condition that the men had to roll him and rub him in the snow for some time before allowing him to approach the fire. But Umentucken was saved, and he became heroic in her eyes. Whether it was the glory acquired by the gallant act just recorded, or whether our hero had now arrived at an age when the tender passion has strongest sway, the writer is unprepared to affirm: for your mountain man is shy of revealing his past gallantries, but from this time on, there are evidences of considerable susceptibility to the charms of the dusky beauties of the mountains and the plains.

The cold of this winter was very severe, insomuch that men and mules were frozen to death. "The frost," says Meek, "used to hang from the roofs of our lodges in the morning, on first waking, in skeins two feet long, and our blankets and whiskers were white with it. But we trappers laid still, and called the camp-keepers to make a fire, and in our close lodges it was soon warm enough.

"The Indians suffered very much. Fuel war scarce on the Snake River, and but little fire could be afforded—just sufficient for the children and their mothers to get warm by, for the fire was fed only with buffalo fat torn in strips, which blazed up quickly and did not last long. Many a time, I have stood off, looking at the fire, but not venturing to approach, when a chief would say, 'Are you cold, my friend? come to the fire'—so kind are these Nez Perces and Flatheads."

The cold was not the only enemy in camp that winter, but famine threatened them. The buffalo had been early driven east of the mountains, and other game was scarce. Sometimes a party of hunters were absent for days, even weeks, without finding more game than would subsist themselves. As the trappers were all hunters in the winter, it frequently happened that Meek and one or more of his associates went on a hunt in company, for the benefit of the camp, which was very hungry at times.

On one of these hunting expeditions that winter, the party consisting of Meek, Hawkins, Doughty, and Antoine Claymore, they had been out nearly a fortnight without killing anything of consequence, and had clambered up the side of the mountains on the frozen snow, in hopes of finding some mountain sheep. As they traveled along under a projecting ledge of rocks, they came to a place where there were the impressions in the snow of enormous grizzly bear feet. Close by was an opening in the rocks, revealing a cavern, and to this, the tracks in the snow conducted. Evidently, the creature had come out of its winter den, and made just one circuit back again. At these signs of game the hunters hesitated—certain it was there, but doubtful how to obtain it.

At length, Doughty proposed to get up on the rocks above the mouth of the cavern and shoot the bear as he came out, if somebody would go in and dislodge him.

"I'm your man," answered Meek.

"And I, too," said Claymore.

"I'll be d—d if we are not as brave as you are," said Hawkins, as he prepared to follow.

On entering the cave, which was sixteen or twenty feet square, and high enough to stand erect in, instead of one, three bears were discovered. They were standing, the largest one in the middle, with their eyes staring at the entrance, but quite quiet, greeting the hunters only with a low growl. Finding that there was a bear apiece to be disposed of, the hunters kept close to the wall, and out of the stream of light from the entrance, while they advanced a little way, cautiously, toward their game, which, however, seemed to take no notice of them. After maneuvering a few minutes to get nearer, Meek finally struck the large bear on the head with his wiping-stick, when it immediately moved off and ran out of the cave. As it came out, Doughty shot, but only wounded it, and it came rushing back, snorting, and running around in a circle, till the well-directed shots from all three killed it on the spot. Two more bears now remained to be disposed of.

The successful shot put Hawkins in high spirits. He began to hallo and laugh, dancing around, and with the others, striking the next largest bear to make him run out, which he soon did, and was shot by Doughty. By this time their guns were reloaded, the men growing more and more elated, and Hawkins declaring they were *all Daniels in the lions' den, and no mistake.* This, and similar expressions, he constantly vociferated, while they drove out the third and smallest bear. As it reached the cave's mouth, three simultaneous shots put an end to the last one, when Hawkins's excitement knew no bounds. "Daniel was a humbug," said he. "Daniel in the lions' den! Of course, it was winter, and the lions were sucking their paws! Tell me no more of Daniel's exploits. We are as good Daniels as he ever dared to be. Hurrah for these Daniels!" With these expressions, and playing many antics by way of rejoicing, the delighted Hawkins finally danced himself out of his *lion's den,* and set to work with the others to prepare for a return to camp.

Sleds were soon constructed out of the branches of the mountain willow, and on these light vehicles the fortunate find of bear meat was soon conveyed to the hungry camp in the plain below. And ever after this singular exploit of the party, Hawkins continued to aver, in language more strong than elegant, that the Scripture Daniel was a humbug compared to himself, and Meek, and Claymore.

# 8

## 1833

In the spring, the camp was visited by a party of twenty Blackfeet, who drove off most of the horses, and among the stolen ones, Bridger's favorite race-horse, Grohean, a Camanche steed of great speed and endurance. To retake the horses, and if possible punish the thieves, a company of the gamest trappers, thirty in number, including Meek, and Kit Carson, who not long before had joined the Rocky Mountain Company, was dispatched on their trail. They had not traveled long before they came up with the Blackfeet, but the horses were nowhere to be seen, having been secreted, after the manner of these thieves, in some defile of the mountains, until the skirmish was over which they knew well enough to anticipate. Accordingly, when the trappers came up, the wily savages were prepared for them. Their numbers were inferior to that of the whites, and accordingly, they assumed an innocent and peace-desiring air, while their head man advanced with the inevitable peace pipe, to have a *talk*. But as their talk was a tissue of lies, the trappers soon lost patience, and a quarrel quickly arose. The Indians betook themselves to the defenses which were selected beforehand, and a fight began, which without giving to either party the victory of arms, ended in the killing of two or three of the Blackfeet, and the wounding very severely of Kit Carson. The firing ceased with nightfall, and when morning came, as usual, the Blackfeet were gone, and the trappers returned to camp without their horses.

The lost animals were soon replaced by purchase from the Nez Perces, and the company divided up into brigades, some destined for the country east of the mountains, and others for the south and west. In this year, Meek rose a grade above the hired trapper, and became one of the order-denominated skin trappers. These, like the hired trappers, depend upon the company to furnish them an outfit, but do not receive regular wages, as do the others. They trap for themselves, only agreeing to sell their beaver to the company which furnishes the outfit, and to no other. In this capacity, our Joe, and a few associates, hunted this spring, in the Snake River and Salt Lake countries, returning as usual to the annual rendezvous, which was appointed this summer to meet on Green River. Here were the Rocky Mountain and American companies, the St. Louis Company, under Capt. WM. Sublette and his friend Campbell, the usual camp of Indian allies, and a few miles distant, that of Captain Bonneville. In addition to all these, was a small company belonging to Captain Stuart, an Englishman of noble family, who was traveling in the far west only to gratify his own love of wild adventure, and admiration of all that is grand and magnificent in nature. With him was an artist named Miller, and several servants, but he usually traveled in company with one or another of the fur companies, thus enjoying their protection, and at the same time, gaining a knowledge of the habits of mountain life.

The rendezvous, at this time, furnished him a striking example of some of the ways of mountain men, least to their honorable fame, and we fear we must confess that our friend Joe Meek, who had been gathering laurels as a valiant hunter and trapper during the three or four years of his apprenticeship, was also becoming fitted, by frequent practice, to graduate in some of the vices of camp life, especially the one of conviviality during rendezvous. Had he not given his permission, we should not perhaps have said what he says of himself, that he was at such times often very *powerful drunk.*

During the indulgence of these excesses, while at this rendezvous, there occurred one of those incidents of wilderness life which make the blood creep with horror. Twelve of the men were bitten by a mad wolf, which hung about the camp for two or three nights. Two of these were seized with madness in camp, some time afterward, and ran off into the mountains, where they perished. One was attacked by the paroxysm while on a hunt, when, throwing himself off his horse, he struggled and foamed at the mouth, gnashing his teeth and barking like a wolf. Yet he retained consciousness enough to warn away his companions, who hastened in search of assistance, but when they returned, he was nowhere to be found. It was thought that he was seen a day or two afterward, but no one could come up with him, and of course, he, too, perished. Another died on his journey to St. Louis and several died at different times within the next two years.

At the time, however, immediately following the visit of the wolf to camp, Captain Stuart was admonishing Meek on the folly of his ways, telling him that the wolf might easily have bitten him, he was so drunk.

"It would have killed him—sure, if it hadn't *cured* him!" said Meek[1]—alluding to the belief that alcohol is a remedy for the poison of hydrophobia.

When sobriety returned, and work was once more to be resumed, Meek returned with three or four associates to the Salt Lake country to trap on the numerous streams that flow down from the mountains to the east of Salt Lake. He had not been long in this region when he fell in on Bear River with a company of Bonneville's men, one hundred and eighteen in number, under Jo Walker, who had been sent to explore the Great Salt Lake, and the adjacent country, to make charts, keep a journal, and in short, make a thorough discovery of all that region. Great expectations were cherished by the captain concerning this favorite expedition, which were, however, utterly blighted, as his historian has recorded. The disappointment and loss which Bonneville suffered from it, gave a tinge of prejudice to his delineations of the trapper's character. It was true that they did not explore Salt Lake, and that they made a long and expensive journey, collecting but few peltries. It is true also, that they caroused in true mountain style, while among the Californians: but that the expedition was unprofitable was due chiefly to the difficul-

---

1. Though it may be that Meek embellishes his tales and sometimes improves his own lines, the fine remark "It would have killed him, sure, if it hadn't cured him!" is confirmed in Edward Warren, by Stewart himself, who sketches this rendezvous in detail.

attending the exploration of a new country, a large portion of which was desert and mountain.[2]

But let us not anticipate. When Meek and his companions fell in with Jo Walker and his company, they resolved to accompany the expedition, for it was *a feather in a man's cap*, and made his services doubly valuable to have become acquainted with a new country, and fitted himself for a pilot.

On leaving Bear River, where the hunters took the precaution to lay in a store of dried meat, the company passed down on the west side of Salt Lake, and found themselves in the Salt Lake desert, where their store, insufficiently large, soon became reduced to almost nothing. Here was experienced again the sufferings to which Meek had once before been subjected in the Digger country, which, in fact, bounded this desert on the northwest. "There was," says Bonneville, "neither tree, nor herbage, nor spring, nor pool, nor running stream, and nothing but parched wastes of sand, where horse and rider were in danger of perishing." Many an emigrant has since confirmed the truth of this account.

It could not be expected that men would continue on in such a country, in that direction which offered no change for the

---

2. Compare Meek's account of the Walker expedition to California with Leonard's, p. 64 ff. Walker followed what was to become one version of the California Trail: Around the northern edge of Salt Lake, to the Humboldt, down that river to its sink, across to Carson Lake. Then across the Sierra Nevada not by what was to be called Donner Pass, the main route of the California Trail, but up a tributary of the East Walker River and across the summits to the headwaters of the Merced and Tuolomne Rivers—according to Ardis M. Walker in the Hafen series, V, p. 367—and into the San Joaquin Valley through what is now Yosemite Park. Walker returned via Walker Pass through the Sierra, up the Owens Valley to his back trail, then along it eastward.

better. Discerning at last a snowy range to the northwest, they traveled in that direction, pinched with famine, and with tongues swollen out of their mouths with thirst. They came at last to a small stream, into which both men and animals plunged to quench their raging thirst.

The instinct of a mule on these desert journeys is something wonderful. We have heard it related by others besides the mountain men, that they will detect the neighborhood of water long before their riders have discovered a sign, and setting up a gallop, when before they could hardly walk, will dash into the water up to their necks, drinking in the life-saving moisture through every pore of the skin, while they prudently refrain from swallowing much of it. If one of a company has been off on a hunt for water, and on finding it has let his mule, drink, when he returns to camp, the other animals will gather about it, and snuff its breath, and even its body, betraying the liveliest interest and envy. It is easy to imagine that in the case of Jo Walker's company, not only the animals but the men were eager to steep themselves in the reviving waters of the first stream which they found on the border of this weary desert.

It proved to be a tributary of Mary's or Ogden's River, along which the company pursued their way, trapping as they went, and living upon the flesh of the beaver. They had now entered upon the same country inhabited by Digger Indians, in which Milton Sublette's brigade had so nearly perished with famine the previous year. It was unexplored, and the natives were as curious about the movements of their white visitors, as Indians always are on the first appearance of civilized men.

They hung about the camps, offering no offenses by day, but contriving to do a great deal of thieving during the nighttime.

Each day, for several days, their numbers increased, until the army which dogged the trappers by day, and filched from them at night, numbered nearly a thousand. They had no guns, but carried clubs, and some bows and arrows. The trappers at length became uneasy at this accumulation of force, even though they had no firearms, for was it not this very style of people, armed with clubs, that attacked Smith's party on the Umpqua, and killed all but four?

"We must kill a lot of them, boys," said Jo Walker. "It will never do to let that crowd get into camp." Accordingly, as the Indians crowded round at a ford of Mary's River, always a favorite time of attack with the savages, Walker gave the order to fire, and the whole company poured a volley into the jostling crowd. The effect was terrible. Seventy-five Diggers bit the dust, while the others, seized with terror and horror at this new and instantaneous mode of death, fled howling away, the trappers pursuing them until satisfied that they were too much frightened to return. This seemed to Captain Bonneville, when he came to hear of it, like an unnecessary and ferocious act. But Bonneville was not an experienced Indian fighter. His views of their character were much governed by his knowledge of the Flatheads and Nez Perces, and also by the immunity from harm he enjoyed among the Shoshonies on the Snake River, where the Hudson's Bay Company had brought them into subjection, and where even two men might travel in safety at the time of his residence in that country.

Walker's company continued on down to the main or Humboldt River, trapping as they went, both for the furs, and for something to eat, and expecting to find that the river whose course they were following through these barren plains, would lead them to some more important river, or to some large lake or inland sea. This was a country entirely unknown, even to the adventurous traders and trappers of the fur companies, who avoided it because it was out of the buffalo range, and because the borders of it, along which they sometimes skirted, were found to be wanting in water-courses in which beaver might be looked for. Walker's company, therefore, was now determined to prosecute their explorations until they came to some new and profitable beaver grounds.

But after a long march through an inhospitable country, they came at last to where the Humboldt sinks itself in a great swampy lake, in the midst of deserts of sage-brush. Here was the end of their great expectations. To the west of them, however, and not far off, rose the lofty summits of the Sierra Nevada range, some of whose peaks were covered with eternal snows. Since they had already made an unprofitable business of their expedition, and failed in its principal aim, that of exploring Salt Lake, they resolved upon crossing the mountains into California, and seeking new fields of adventure on the western side of the Nevada mountains.

Accordingly, although it was already late in the autumn, the party pushed on toward the west, until they came to Pyramid Lake, another of those swampy lakes which are frequently met with near the eastern base of these Sierras. Into this flowed a stream similar to the Humboldt, which came from the south, and they believed, had its rise in the mountains. As it was important to find a good pass, they took their course along this stream, which they named Trucker's River, and continued along it to its headwaters in the Sierras.

And now began the arduous labor of crossing an unknown range of lofty mountains. Mountaineers as they were, they found it a difficult undertaking, and one attended with considerable peril. For a period of more than three weeks they were struggling with these dangers, hunting paths for their mules and horses, traveling around canyons thousands of feet deep, sometimes sinking in new-fallen snow, always hungry, and often in peril from starvation. Sometimes, they scrambled up almost smooth declivities of granite, that offered no foothold save the occasional seams in the rock, at others they traveled through pine forests made nearly impassable by snow, and at other times on a ridge which wind and sun made bare for them. All around, rose rocky peaks and pinnacles fretted by ages of denudation to very spears and needles of a burned-looking, red-colored rock. Below, were spread out immense fields, or rather oceans, of granite that seemed once to have been a molten sea, whose waves were suddenly congealed. From the fissures between these billows grew stunted pines, which had found a scanty soil far down in the crevices of the rock for their hardy roots. Following the course of any stream flowing in the right direction for their purpose, they came not infrequently to some small fertile valley, set in amid the rocks like a cup, and often containing in its depth a bright little lake. These are the oases in the mountain deserts. But the lateness of the season made it necessary to avoid the high valleys on account of the snow, which in winter accumulates to a depth of twenty feet.

Great was the exultation of the mountaineers when they emerged from the toils and dangers, safe into the bright and sunny plains of California, having explored almost the identical route since fixed upon for the Union Pacific Railroad.[3]

They proceeded down the Sacramento valley, toward the coast, after recruiting their horses on the ripe wild oats, and the freshly springing grass which the December rains had started into life, and themselves on the plentiful game of the foothills. Something of the stimulus of the Californian climate seemed to be imparted to the ever-buoyant blood of these hardy and danger-despising men. They were mad with delight on finding themselves, after crossing the stern Sierras, in a land of sunshine and plenty, a beautiful land of verdant hills and tawny plains, of streams winding between rows of alder and willow, and valleys dotted with picturesque groves of the evergreen oak. Instead of the wild blasts which they were used to encounter in December, they experienced here only those dainty and wooing airs which poets have ascribed to spring, but which seldom come even with the last May days in an eastern climate.

In the San José valley, they encountered a party of one hundred soldiers, which the Spanish government at Monterey had sent out to take a party of Indians accused of stealing cattle. The soldiers were native Californians, descendants of the mixed blood of Spain and Mexico, a wild, jaunty-looking set of fellows, who at first were inclined to take Walker's party for a band of cattle thieves, and to march them off to Monterey. But the Rocky Mountain trapper was not likely to be taken prisoner by any such brigade as the dashing *cabelleros* of Monterey.

---

3. Meek gives a different version of the route, claiming that he crossed the mountains via Donner Pass, pioneering the route of the future California Trail and the future Union Pacific Railroad. Perhaps he and the other free trappers separated temporarily from the main Walker party, as they were free to do. Or perhaps—less likely—Joe misremembered in later years.

After astonishing them with a series of whoops and yells, and trying to astonish them with feats of horsemanship, they began to discover that when it came to the latter accomplishment, even mountain men could learn something from a native Californian. In this latter frame of mind, they consented to be conducted to Monterey as prisoners or not, just as the Spanish government should hereafter be pleased to decree, and they had confidence in themselves that they should be able to bend that high and mighty authority to their own purposes thereafter.

Nor were they mistaken in their calculations. Their fearless, free and easy style, united to their complete furnishing of arms, their numbers, and their superior ability to stand up under the demoralizing effect of the favorite *aguadiente*, soon so far influenced the soldiery at least, that the trappers were allowed perfect freedom under the very eyes of the jealous Spanish government, and were treated with all hospitality.

The month which the trappers spent at Monterey was their *red letter day* for a long time after. The habits of the Californians accorded with their own, with just difference enough to furnish them with novelties and excitements such as gave a zest to their intercourse. The Californian, and the mountain men, were alike centaurs. Horses were their necessity, and their delight, and the plains swarmed with them, as also with wild cattle, descendants of those imported by the Jesuit Fathers in the early days of the Missions. These horses and cattle were placed at the will and pleasure of the trappers. They feasted on one, and bestrode the other as it suited them. They attended bull-fights, ran races, threw the lasso, and played monte, with a relish that delighted the inhabitants of Monterey.

The partial civilization of the Californians accorded with

every feeling to which the mountain men could be brought to confess. To them, the refinements of cities would have been oppressive. The adobe houses of Monterey were not so restraining in their elegance as to trouble the sensations of men used to the heavens for a roof in summer, and a skin lodge for shelter in winter. Some fruits and vegetables, articles not tasted for years, they obtained at the missions, where the priests received them courteously and hospitably, as they had done Jedediah Smith and his company, five years before, when on their long and disastrous journey, they found themselves almost destitute of the necessaries of life, upon their arrival in California. There was something, too, in the dress of the people, both men and women, which agreed with, while differing from, the dress of the mountaineers and their now absent Indian dulcineas.

The men wore garments of many colors, consisting of blue velveteen breeches and jacket, the jacket having a scarlet collar and cuffs, and the breeches being open at the knee to display the stocking of white. Beneath these were displayed high buskins made of deer skin, fringed down the outside of the ankle, and laced with a cord and tassels. On the head was worn a broad-brimmed *sombrero,* and over the shoulders the jaunty Mexican *sarape*. When they rode, the Californians wore enormous spurs, fastened on by jingling chains. Their saddles were so shaped that it was difficult to dislodge the rider, being high before and behind, and the indispensable lasso hung coiled from the pommel. Their stirrups were of wood, broad on the bottom, with a guard of leather that protected the fancy buskin of the horseman from injury. Thus accoutred, and mounted on a wild horse, the Californian was a suitable comrade, in appearance, at least, for the buckskin-clad trapper, with his high beaver-skin cap, his gay scarf, and moccasins, and profusion of arms.

The dress of the women was a gown of gaudy calico or silk, and a bright-colored shawl, which served for mantilla and bonnet together. They were well formed, with languishing eyes and soft voices, and doubtless appeared charming in the eyes of our band of trappers, with whom they associated freely at fandangoes, bull-fights, or bear-baitings. In such company, what wonder that Bonneville's men lingered for a whole month! What wonder that the California expedition was a favorite theme by campfires, for a long time subsequent?

---

**1834**

In February, the trappers bethought themselves of returning to the mountains. The route fixed upon was one which should take them through Southern California, and New Mexico, along the course of all the principal rivers. Crossing the coast mountains into the valley of the San Joaquin, they followed its windings until they came to its rise in the Lulare Lake. Thence turning in a southeasterly course, they came to the Colorado, at the Mohave villages, where they traded with the natives, whom they found friendly. Keeping on down the Colorado, to the mouth of the Gila, they turned back from that river, and ascended the Colorado once more, to Williams's Fork, and up the latter stream to some distance, when they fell in with a company of sixty men under Frapp and Jervais, two of the partners in the Rocky Mountain Company. The meeting was joyful on all sides, but particularly so between Meek and some of his old comrades, with whom he had fought Indians and grizzly bears, or set beaver traps on some lonely stream in the Blackfoot country. A lively exchange of questions and answers took place, while gaiety and good feeling reigned.

Frapp had been out quite as long as the Monterey party. It was seldom that the brigade which traversed the southern country, on the Colorado, and its large tributaries, returned to winter quarters, for in the region where they trapped winter was unknown, and the journey to the northern country a long and hazardous one. But the reunited trappers had each their own experiences to relate.

The two companies united and made a party nearly two hundred strong. Keeping with Frapp, they crossed over from Williams's Fork to the Colorado Chiquito river, at the Moquis village, where some of the men disgraced themselves far more than did Jo Walker's party at the crossing of Mary's River. For the Moquis were a half-civilized nation, who had houses and gardens, and conducted themselves kindly, or at the worst peaceably, toward properly behaved strangers. These trappers, instead of approaching them with offers of purchase, lawlessly entered their gardens, rifling them of whatever fruit or melons were ripe, and not hesitating to destroy that which was not ripe. To this, as might be expected, the Moquises objected and were shot down for so doing. In this truly infamous affair, fifteen or twenty of them were killed.

"I didn't belong to that crowd," says Joe Meek. "I sat on the fence and saw it, though. It was a shameful thing."

From the Moquis village, the joint companies crossed the country in a northeasterly direction, crossing several branches of the Colorado at their headwaters, which of course finally brought them to the headwaters of the Rio Grande. The journey from the mouth of the Gila, though long, extended over a country comparatively safe. Either farther to the south or east, the caravan would have been in danger of a raid from the most dangerous tribes on the continent.[4]

---

4. Meek indicates that the Walker party returned to the mountains by a route through the southwest, which is not true. Perhaps Joe and some of the free trappers went their own way. For that possibility Joe has corroboration: He says they met Frapp and Jervais on the Williams Fork of the Colorado that spring, and LeRoy Hafen points out—the Hafen series, III, p. 135—that a letter of Tom Fitzpatrick's confirms that Frapp and Jervais were there that spring. The return route through the Southwest: Across the Sierra from Tulare Lake by an unspecified route. Across the Mojave Desert by an unspecified route to the Mojave Villages—near modern Needles, California. Down the Colorado to the mouth of the Gila River, back up the Colorado to the Williams Fork and up that river, across to the Little Colorado River, to the village of a pueblo people—perhaps the Hopi—then northeasterly across several branches of the Colorado at their headwaters—evidently the branches that flow together to make the San Juan River—to the headwaters of the Rio Grande. Ultimately Joe then went to the headwaters of the Colorado—Grande—River, north to the sources of the North Platte in New or North Park, to Old or Middle Park, to the Little Snake River, and to the rendezvous on Ham's Fork of the Green—see p. 158. But first he claims a detour, discussed under the next chapter.

# 9

## 1834

But Joe Meek was not destined to return to the Rocky Mountains without having had an Indian fight. If adventures did not come in his way he was the man to put himself in the way of adventures.

While the camp was on its way from the neighborhood of Grande River to the New Park, Meek, Kit Carson, and Mitchell, with three Delaware Indians, named Tom Hill, Manhead, and Jonas, went on a hunt across to the east of Grande River, in the country lying between the Arkansas and Cimarron, where numerous small branches of these rivers head together, or within a small extent of country.

They were about one hundred and fifty miles from camp, and traveling across the open plain between the streams, one beautiful May morning, when about five miles off, they descried a large band of Indians mounted, and galloping toward them. As they were in the Camanche country, they knew what to expect if they allowed themselves to be taken prisoners. They gave but a moment to the observation of their foes, but that one moment revealed a spirited scene. Fully two hundred Camanches[1], their warriors in front, large and well-formed men, mounted on fleet and powerful horses, armed with spears and battle-axes, racing like the wind over the prairie, their feather head-dresses bending to the breeze, that swept past them in the race with double force, all distinctly seen in the clear air of the prairie, and giving the beholder a thrill of fear mingled with admiration.

The first moment given to this spectacle, the second one was employed to devise some means of escape. To run was useless. The swift Camanche steeds would soon overtake them, and then their horrible doom was fixed. No covert was at hand, neither thicket nor ravine, as in the mountains there might have been. Carson and Meek exchanged two or three sentences. At last, "We must kill our mules!" said they.

That seems a strange devise to the uninitiated reader, who no doubt believes that in such a case, their mules must be their salvation. And so they were intended to be. In this plight, a dead mule was far more useful than a live one. To the ground sprang every man, and placing their mules, seven in number, in a ring,

---

1. This battle with the Comanches has been sometimes thought invented, sometimes thought factual. Harvey E. Tobie discreetly omits it from his biographical sketch of Joe in the Hafen series—I, p. 313-335. In his sketch of Carson's life, though, Harvey Carter argues that the incident did take place—Hafen series, VI, p. 108—but that it must have been at a time other than Meek assigns it. Vestal argues for the incident at length, citing verbal confirmation from George Bent—Vestal, note, p. 148-149. Vestal reminds us that Joe's claim of 42 Comanche dead is likely an exaggeration. It seems to the editor that Meek has enough corroboration for the basic outline of the incident to be accepted.

they, in an instant cut their throats with their hunting knives, and held on to the bridles until each animal fell dead in its appointed place. Then hastily scooping up what earth they could with knives, they made themselves a fort—a hole to stand in for each man, and a dead mule for a breastwork.

In less than half an hour the Camanches charged on them, with the medicine man in advance shouting, gesticulating, and making a desperate clatter with a rattle which he carried and shook violently. The yelling, the whooping, the rattling, the force of the charge were appalling. But the little garrison in the mule fort did not waver. The Camanche horses did. They could not be made to charge upon the bloody carcasses of the mules, nor near enough for their riders to throw a spear into the fort.

This was what the trappers had relied upon. They were cool and determined, while terribly excited and wrought up by their situation. It was agreed that no more than three should fire at a time, the other three reserving their fire while the empty guns could be reloaded. They were to pick their men, and kill one at every shot.

They acted up to their regulations. At the charge the Camanche horses recoiled and could not be urged upon the fort of slaughtered mules. The three whites fired first, and the medicine man and two other Camanches fell. When a medicine man is killed, the others retire to hold a council and appoint another, for without their *medicine*, they could not expect success in battle. This was time gained. The warriors retired, while their women came up and carried off the dead.

After devoting a little time to bewailing the departed, another chief was appointed to the head place, and another furious charge was made with the same results as before. Three more warriors bit the dust, while the spears of their brethren, attached to long hair ropes by which they could be withdrawn, fell short of reaching the men in the fort. Again and again, the Camanches made a fruitless charge, losing, as often as they repeated it, three warriors, either dead or wounded. Three times that day, the head chief or medicine man was killed, and when that happened, the heroes in the fort got a little time to breathe. While the warriors held a council, the women took care of the wounded and slain.

As the women approached the fort to carry off the fallen warriors, they mocked and reviled the little band of trappers, calling them *women*, for fighting in a fort, and resorting to the usual Indian ridicule and gasconade. Occasionally, also, a warrior raced at full speed past the fort apparently to take observations. Thus the battle continued through the entire day.

It was terrible work for the trappers. The burning sun of the plains shone on them, scorching them to faintness. Their faces were begrimed with powder and dust, their throats parched, and tongues swollen with thirst, and their whole frames aching from their cramped positions, as well as the excitement and fatigue of the battle. But they dared not relax their vigilance for a moment. They were fighting for their lives, and they meant to win.

At length, the sunset on that bloody and wearisome day. Forty-two Camanches were killed, and several more wounded, for the charge had been repeated fifteen or twenty times. The Indians drew off at nightfall to mourn over their dead, and hold a council. Probably they had lost faith in their medicines, or believed that the trappers possessed one far greater than any of theirs. Under the friendly cover of the night, the six heroes who had fought successfully more than a hundred Camanches, took each his blanket and his gun, and bidding a brief adieu to dead mules and beaver packs, set out to return to camp.

When a mountain man had a journey to perform on foot, to travel express, or to escape from an enemy, he fell into what is called a dog trot, and ran in that manner, sometimes, all day. On the present occasion, the six, escaping for life, ran all night, and found no water for seventy-five miles. When they did at last come to a clear running stream, their thankfulness was equal to their necessity. "For," says Meek, "thirst is the greatest suffering I ever experienced. It is far worse than hunger or pain."

Having rested and refreshed themselves at the stream, they kept on without much delay until they reached camp in that beautiful valley of the Rocky Mountains called the New, or the South Park.

While they remained in the South Park, Mr. Guthrie, one of the Rocky Mountain Company's traders, was killed by lightning. A number of persons were collected in the lodge of the Booshway, Frapp, to avoid the rising tempest, when Guthrie, who was leaning against the lodge pole, was struck by a flash of the electric current, and fell dead instantly. Frapp rushed out of the lodge, partly bewildered himself by the shock, and under the impression that Guthrie had been shot. Frapp was a German, and spoke English somewhat imperfectly. In the excitement of the moment, he shouted out, "Py Gott, who did shoot Guttery!"

"G—a' Mighty, I expect: He's a firing into camp," drawled out Hawkins, whose ready wit was very disregardful of sacred names and subjects.

The mountaineers were familiar with the most awful aspects of nature, and if their familiarity had not bred contempt, it had at least hardened them to those solemn impressions which other men would have felt under their influence.

From New Park[2], Meek traveled north with the main camp, passing first to the Old Park, thence to the Little Snake, a branch of Bear River, thence to Pilot Butte, and finally to Green River to rendezvous, having traveled in the past year about three thousand miles, on horseback, through new and often dangerous countries. It is easy to believe that the Monterey expedition was the popular theme in camp during rendezvous. It had been difficult to get volunteers for Bonneville's Salt Lake Exploration: but such was the wild adventure to which it led, that volunteering for a trip to Monterey would have been exceedingly popular immediately thereafter.

On Bear River, Bonneville's men fell in with their commander, Captain Bonneville, whose disappointment and indignation at the failure of his plans was exceedingly great. In this indignation, there was considerable justice, yet much of his disappointment was owing to causes which a more experienced trader would have avoided. The only conclusion which can be arrived at by an impartial observer of the events of 1832–35, is that none but certain men of long experience and liberal means, could succeed in the business of the fur trade. There were too many chances of loss, too many wild elements to be mingled in amity, and too powerful opposition from the old established companies. Captain Bonneville's experience was no different from Mr. Wyeth's. In both cases, there was much effort, outlay, and loss.

---

2. New Park is North Park, not South Park. Carson says the incident of Guthrie's being shot down by God occurred in South Park, also called Bayou Salade.

Nor was their failure owing to any action of the Hudson's Bay Company, different from, or more tyrannical, than the action of the American companies, as has frequently been represented. It was the American companies in the Rocky Mountains that drove both Bonneville and Wyeth out of the field. Their inexperience could not cope with the thorough knowledge of the business, and the country, which their older rivals possessed. Raw recruits were no match, in trapping or fighting, for old mountaineers: and those veterans who had served long under certain leaders could not be inveigled from their service except upon the most extravagant offers—and these extravagant wages, which if one paid, the other must, would not allow a profit to either of the rivals.

"How much does your company pay you?" asked Bonneville of Meek, to whom he was complaining of the conduct of his men on the Monterey expedition.

"Fifteen hundred dollars," answered Meek.

"Yes, and *I* will give it to you," said Bonneville with bitterness.

It was quite true. Such was the competition aroused by the captain's efforts to secure good men and pilots, that rather than lose them to a rival company, the Rocky Mountain Company paid a few of their best men the wages above named.

# 10

## 1834

The gossip at rendezvous was this year of an unusually exciting character. Of the brigades which left for different parts of the country the previous summer, the Monterey travelers were not the only ones who had met with adventures. Fitzpatrick, who had led a party into the Crow country that autumn, had met with a characteristic reception from that nation of cunning vagabonds.

Being with his party on Lougue River, in the early part of September, he discovered that he was being dogged by a considerable band of Crows, and endeavored to elude their spying, but all to no purpose. The Crow chief kept in his neighborhood, and finally expressed a desire to bring his camp alongside that of Fitzpatrick, pretending to have the most friendly and honorable sentiments toward his white neighbors. But not feeling any confidence in Crow friendship, Fitzpatrick declined, and moved camp a few miles away. Not, however, wishing to offend the dignity of the apparently friendly chief, he took a small escort, and went to pay a visit to his Crow neighbors, so that they might see that he was not afraid to trust them. Alas, vain subterfuge!

While he was exchanging civilities with the Crow chief, a party of the young braves stole out of camp, and taking advantage of the leader's absence, made an attack on his camp, so sudden and successful that not a horse, nor anything else which they could make booty of was left. Even Captain Stuart, who was traveling with Fitzpatrick, and who was an active officer, was powerless to resist the attack, and had to consent to see the camp rifled of everything valuable.

In the meantime, Fitzpatrick, after concluding his visit in the most amicable manner, was returning to camp, when he was met by the exultant braves, who added insult to injury by robbing him of his horse, gun, and nearly all his clothes, leaving him to return to his party in a deplorable condition, to the great amusement of the trappers, and his own chagrin.[1]

However, the next day, a talk was held with the head chief of the Crows, to whom Fitzpatrick represented the infamy of such treacherous conduct in a very strong light. In answer to this reproof, the chief disowned all knowledge of the affair, saying that he could not always control the conduct of the young men, who would be a little wild now and then, in spite of the best Crow precepts, but that he would do what he could to have the

---

1. See Irving, p. 207-208, for a basic account of this robbery. DeVoto—p. 124-131—lays the blame squarely at the feet of American fur in general and Jim Beckwourth in particular. Beckwourth—p. 274-282—says he exerted himself to save the white men's lives and help them get some of their property back, but he has not been generally believed.

property restored. Accordingly, after more talk, and much eloquence on the part of Fitzpatrick, the chief part of the plunder was returned to him, including the horses and rifles of the men, together with a little ammunition, and a few beaver traps.

Fitzpatrick understood the meaning of this apparent fairness, and hastened to get out of the Crow country before another raid by the mischievous young braves, at a time when their chief was not *honor bound*, should deprive him of the recovered property. That his conjecture was well founded, was proven by the numerous petty thefts which were committed, and by the loss of several horses and mules, before he could remove them beyond the limits of the Crow territory.

While the trappers exchanged accounts of their individual experiences, the leaders had more important matters to gossip over. The rivalry between the several fur companies was now at its climax. Through the energy and ability of Captain Sublette of the St. Louis Company, and the experience and industry of the Rocky Mountain Company, which Captain Sublette still continued to control in a measure, the power still remained with them. The American Company had never been able to cope with them in the Rocky Mountains, and the St. Louis Company were already invading their territory on the Missouri River, by carrying goods up that river in boats, to trade with the Indians under the very walls of the American Company's forts.

In August of the previous year, when Mr. Nathaniel Wyeth had started on his return to the States, he was accompanied as far as the mouth of the Yellowstone by Milton Sublette, and had engaged with that gentleman to furnish him with goods the following year, as he believed he could do, cheaper than the St.

Louis Company, who purchased their goods in St. Louis at a great advance on Boston prices.[2] But Milton Sublette fell in with his brother, the captain, at the mouth of the Yellowstone, with a keel-boat loaded with merchandise, and while Wyeth pursued his way eastward to purchase the Indian goods which were intended to supply the wants of the fur traders in the Rocky Mountains, at a profit to him, and an advantage to them, the captain was persuading his brother not to encourage any interlopers in the Indian trade, but to continue to buy goods from himself, as formerly. So potent were his arguments, that Milton yielded to them, in spite of his engagement with Wyeth. Thus, during the autumn of 1833, while Bonneville was being wronged and robbed, as he afterward became convinced, by his men under Walker, and anticipated in the hunting ground selected for himself, in the Crow country, by Fitzpatrick, as he had previously been in the Snake country by Milton Sublette, Wyeth was proceeding to Boston in good faith, to execute what proved to be a fool's errand. Bonneville also had gone on another, when after the trapping season was over, he left his camp to winter on the Snake River, and started with a small escort to visit the Columbia, and select a spot for a trading post on the lower portion of that river. On arriving at Walla-Walla[3], after a hard journey over the Blue Mountains in the winter, the agent at that post had refused to supply him with provisions to prosecute his

---

2. William Sublette in fact arrived at rendezvous before Wyeth did, and by pressing the partners of Rocky Mountain Fur Co. for past debts, forced that firm to reorganize under the name Fitzpatrick—Milton—Sublette & Bridger. See the Hafen series, I, p. 143-145.

3. For Irving's similar account of Bonneville's trip to Fort Walla-Walla and back, see Irving, p. 219.

journey, and given him to understand that the Hudson's Bay Company might be polite and hospitable to Captain Bonneville as the gentleman, but that it was against their regulations to encourage the advent of other traders who would interfere with their business, and unsettle the minds of the Indians in that region.

This reply so annoyed the captain, that he refused the well-meant advice of Mr. Pambrun that he should not undertake to recross the Blue Mountains in March snows, but travel under the escort of Mr. Payette, one of the Hudson's Bay Company's leaders, who was about starting for the Nez Percé country by a safer if more circuitous route. He, therefore, set out to return by the route he came, and only arrived at camp in May 1834, after many dangers and difficulties. From the Portneuf River, he then proceeded with his camp to explore the Little Snake River, and Snake Lake, and it was while so doing that he fell in with his men just returned from Monterey.

Such was the relative position of the several fur companies in the Rocky Mountains in 1834, and it was of such matters that the leaders talked in the lodge of the Booshways, at rendezvous. In the meantime Wyeth arrived in the mountains with his goods, as he had contracted with Milton Sublette in the previous year. But on his heels came Captain Sublette, also with goods, and the Rocky Mountain Company violated their contract with Wyeth, and purchased of their old leader.

Thus was Wyeth left, with his goods on his hands, in a country where it was impossible to sell them, and useless to undertake an opposition to the already established fur traders and trappers. His indignation was great and certainly was just. In his interview with the Rocky Mountain Company, in reply to their excuses for, and vindication of their conduct, his answer was:

"Gentlemen, I will roll a stone into your garden that you will never be able to get out."

And he kept his promise, for that same autumn, he moved on to the Snake River, and built Fort Hall, storing his goods therein. The next year, he sold out goods and fort to the Hudson's Bay Company, and the stone was in the garden of the Rocky Mountain Fur Company that they were never able to dislodge. When Wyeth had built his fort and left it in charge of an agent, he dispatched a party of trappers to hunt in the Big Blackfoot country, under Joseph Gale, who had previously been in the service of the Rocky Mountain Company, and of whom we shall learn more hereafter, while he set out for the Columbia to meet his vessel, and establish a salmon fishery. The fate of that enterprise has already been recorded.

As for Bonneville, he made one more effort to reach the Lower Columbia, failing, however, a second time, for the same reason as before—he could not subsist himself and company in a country where even every Indian refused to sell to him either furs or provisions. After being reduced to horseflesh, and finding no encouragement that his condition would be improved farther down the river, he turned back once more from about Wallah-Wallah, and returned to the mountains, and from there to the east in the following year. A company of his trappers, however, continued to hunt for him east of the mountains for two or three years longer.

The rivalry between the Rocky Mountain and American companies was this year diminished by their mutually agreeing to confine themselves to certain parts of the country, which treaty continued for two years, when they united in one company. They were then, with the exception of a few lone traders, the only competitors of the Hudson's Bay Company, for the fur trade of the West.

# 11

## 1834

The Rocky Mountain Company now confined themselves to the country lying east of the mountains, and upon the headwaters and tributaries of the Missouri, a country very productive in furs, and furnishing abundance of game. But it was also the most dangerous of all the northern fur-hunting territory, as it was the home of those two nations of desperadoes, the Crows and Blackfeet. During the two years in which the company may have been said almost to reside there, desperate encounters and hair-breadth escapes were incidents of daily occurrence to some of the numerous trapping parties.

The camp had reached the Blackfoot country in the autumn of this year, and the trappers were out in all directions, hunting beaver in the numerous small streams that flow into the Missouri. On a small branch of the Gallatin Fork, some of the trappers fell in with a party of Wyeth's men, under Joseph Gale.

When their neighborhood became known to the Rocky Mountain camp, Meek and a party of sixteen of his associates immediately resolved to pay them a visit, and inquire into their experience since leaving rendezvous. These visits between different camps are usually seasons of great interest and general rejoicing. But glad as Gale and his men were to meet with old friends, when the first burst of hearty greeting was over, they had but a sorry experience to relate. They had been out a long time. The Blackfeet had used them badly—several men had been killed. Their guns were out of order, their ammunition all but exhausted. They were destitute, or nearly so, of traps, blankets, knives, everything. They were what the Indian and the mountain man call *very poor.*[1]

Half the night was spent in recounting all that had passed in both companies since the fall hunt began. Little sympathy did Wyeth's men receive for their forlorn condition, for sympathy is repudiated by your true mountaineer for himself, nor will he furnish it to others. The absurd and humorous, or the daring and reckless, side of a story is the only one which is dwelled upon in narrating his adventures. The laugh which is raised at his expense when he has a tale of woes to communicate, is a better tonic to his dejected spirits than the gentlest pity would be. Thus, lashed into courage again, he is ready to declare that all his troubles were only so much pastime.

It was this sort of cheer which the trapping party conveyed to Wyeth's men on this visit, and it was gratefully received, as being of the true kind.

---

1. For his similar account of this incident—misdated—see Carson, p. 59-61. Most authorities date this incident 1835 rather than 1834.

In the morning the party set out to return to camp, Meek and Liggit starting in advance of the others. They had not proceeded far when they were fired on by a large band of Blackfeet, who came upon them quite suddenly, and thinking these two trappers easy game, set up a yell and dashed at them. As Meek and Liggit turned back and ran to Gale's camp, the Indians in full chase charged on them, and rushed pell-mell into the midst of camp, almost before they had time to discover that they had surprised so large a party of whites. So sudden was their advent, that they had almost taken the camp before the whites could recover from the confusion of the charge.

It was but a momentary shock, however. In another instant, the roar of twenty guns reverberated from the mountains that rose high on either side of camp. The Blackfeet were taken in a snare, but they rallied and fell back beyond the grove in which the camp was situated, setting on fire the dry grass as they went. The fire quickly spread to the grove, and shot up the pine trees in splendid columns of flame, that seemed to lick the face of heaven. The Indians kept close behind the fire, shooting into camp whenever they could approach near enough, the trappers replying by frequent volleys. The yells of the savages, the noise of the flames roaring in the trees, the bellowing of the guns, whose echoes rolled among the hills, and the excitement of a battle for life, made the scene one long to be remembered with distinctness.

Both sides fought with desperation. The Blackfoot blood was up—the trapper blood, no less. Gale's men, from having no ammunition, nor guns that were in order, could do little more than take charge of the horses, which they led out into the bottomland to escape the fire, fight the flames, and look after the camp goods. The few whose guns were available, showed the game spirit, and the fight became interesting as an exhibition of what mountain white men could do in a contest of one to ten, with the crack warriors of the red race. It was, at any time, a game party, consisting of Meek, Carson, Hawkins, Gale, Liggit, Rider, Robinson, Anderson, Russel, Larison, Ward, Parmaley, Wade, Michael Head, and a few others whose names have been forgotten.

The trappers being driven out of the grove by the fire, were forced to take to the open ground. The Indians, following the fire, had the advantage of the shelter afforded by the trees, and their shots made havoc among the horses, most of which were killed because they could not be taken. As for the trappers, they used the horses for defense, making rifle-pits behind them, when no other covert could be found. In this manner the battle was sustained until three o'clock in the afternoon, without loss of life to the whites, though several men were wounded.

At three in the afternoon, the Blackfoot chief ordered a retreat, calling out to the trappers that they would fight no more. Though their loss had been heavy, they still greatly outnumbered the whites, nor would the condition of the arms and the small amount of ammunition left permit the trappers to pursue them. The Indians were severely beaten, and no longer in a condition to fight, all of which was highly satisfactory to the victors. The only regret was, that Bridger's camp, which had become aware during the day that a battle was going on in the neighborhood, did not arrive early enough to exterminate the whole band. As it was, the big camp only came up in time to assist in taking care of the wounded. The destruction of their horses put an end to the independent existence of Gale's brigade, which joined itself and its fortunes to Bridger's command for the remainder of the year. Had it not been for the fortunate visit of the trappers to Gale's camp, without doubt, every man in it would have perished at the hands of the Blackfeet: a piece of bad fortune not unaccordant with that which seemed to pursue the enterprises set on foot by the active but unlucky New England trader.

Not long after this battle with the Blackfeet, Meek and a trapper named Crow, with two Shawnees, went over into the Crow country to trap on Pryor's River, a branch of the Yellowstone. On coming to the pass in the mountains between the Gallatin Fork of the Missouri and the great bend in the Yellowstone, called Pryor's Gap, Meek rode forward, with the mad-cap spirit strong in him, to *have a little fun with the boys*, and advancing a short distance into the pass, wheeled suddenly, and came racing back, whooping and yelling, to make his comrades think he had discovered Indians. And lo! as if his yells had invoked them from the rocks and trees, a war party suddenly emerged from the pass, on the heels of the jester, and what had been sport speedily became earnest, as the trappers turned their horses' heads and made off in the direction of camp. They had a fine race of it, and heard other yells and war-whoops besides their own, but they contrived to elude their pursuers, returning safely to camp.

This freak of Meek's was, after all, a fortunate inspiration, for had the four trappers entered the pass and come upon the war party of Crows, they would never have escaped alive.

A few days after, the same party set out again, and succeeded in reaching Pryor's River unmolested, and setting their traps. They remained some time in this neighborhood trapping, but the season had become pretty well advanced, and they were thinking of returning to camp for the winter. The Shawnees set out in one direction to take up their traps, Meek and Crow in another. The stream where their traps were set was bordered by thickets of willow, wild cherry, and plum trees, and the bank was about ten feet above the water at this season of the year.

Meek had his traps set in the stream about midway between two thickets. As he approached the river, he observed with the quick eye of an experienced mountain man, certain signs which gave him little satisfaction. The buffalo were moving off as if disturbed. A bear ran suddenly out of its covert among the willows.

"I told Crow," said Meek, "that I didn't like to go in there. He laughed at me, and called me a coward. 'All the same,' I said. I had no fancy for the place just then—I didn't like the indications. But he kept jeering me, and at last, I got mad and started in. Just as I got to my traps, I discovered that two red devils war a watching me from the shelter of the thicket to my left, about two rods off. When they saw that they war discovered they raised their guns and fired. I turned my horse's head at the same instant, and one ball passed through his neck, under the neck bone, and the other through his withers, just forward of my saddle.

"Seeing that they had not hit me, one of them ran up with a spear to spear me. My horse war rearing and pitching from the pain of his wounds, so that I could with difficulty govern him, but I had my gun laid across my arm, and when I fired I killed the rascal with the spear. Up to that moment, I had supposed that them two war all I had to deal with. But as I got my horse turned round, with my arm raised to fire at the other red devil, I encountered the main party, forty-nine of them, who war in the bed of the stream, and had been covered by the bank. They fired a volley at me. Eleven balls passed through my blanket, under my arm, which war raised. I thought it time to run, and run I did. Crow war about two hundred yards off. So quick had all this happened, that he had not stirred from the spot whar I left him. When I came up to him I called out that I must get on behind him, for my horse war sick and staggering.

"'Try him again,' said Crow, who war as anxious to be off as I war. I did try him agin, and sure enough, he got up a gallop, and away we went, the Blackfeet after us. But being mounted, we had the advantage, and soon distanced them. Before we had run a mile, I had to dismount and breathe my horse. We war in a narrow pass whar it war impossible to hide, so when the Indians came up with us, as they did, while I war dismounted we took sure aim and killed the two foremost ones. Before the others could get close enough to fire we war off agin. It didn't take much urging to make my horse go then, for the yells of them Blackfeet spurred him on.

"When we had run another mile I dismounted agin, for fear that my horse would give out, and agin we war overtaken. Them Blackfeet are powerful runners—no better than us mountain men, though. This time we served them just as we did before. We picked off two of the foremost, and then went on, the rest whooping after us. We war overtaken a third time in the same manner, and the third time two Blackfeet fell dead in advance. At this, they took the hint. Six warriors already gone for two white scalps and two horses, they didn't know how many more would go in the same way. And I reckon they had run about all they wanted to, anyway."

It is only necessary to add that Meek and Crow arrived safely at camp, and that the Shawnees came in after a day or two all right. Soon after, the whole command under Bridger moved on to the Yellowstone, and went into winter camp in the great bend of that river, where buffalo were plenty, and cottonwood was in abundance.

---

## 1835

Toward spring, however, the game had nearly all disappeared from the neighborhood of the camp, and the hunters were forced to follow the buffalo in their migration eastward. On one of these expeditions, a party of six trappers, including Meek, and a man named Rose, made their camp on Clarke's Fork of the Yellowstone. The first night in camp Rose had a dream with which he was very much impressed. He dreamed of shaking hands with a large white bear, which insisted on taking his right hand for that friendly ceremony. He had not given it very willingly, for he knew too much about bears in general to desire to be on very intimate terms with them.

Seeing that the dream troubled Rose, who was superstitiously inclined, Meek resorted to that *certain medicine for minds diseased* which was in use in the mountains, and added to the distress of Rose his interpretation, in the spirit of ridicule, telling him that he was an adept in the matter of dreams, and that unless he, Rose, was very mindful of himself that day, he would shake hands with Beelzebub before he slept again.

With this comforting assurance, Rose set out with the remainder of the party to hunt buffalo. They had proceeded about three miles from camp, Rose riding in advance, when they suddenly encountered a company of Blackfeet, nine in number, spies from a war party of one hundred and fifty, that was prowling and marauding through the country on the lookout for small parties from the camp of Bridger. The Blackfeet fired on the party as it came up, from their place of concealment, a ball striking Rose's right arm, and breaking it at the elbow. This caused his gun to fall, and an Indian sprang forward and raised it up quickly, aiming it at Meek. The ball passed through his cap without doing any other harm. By this time the trappers were made aware of an ambuscade, but how numerous the enemy was they could not determine. However, as the rest, who were well mounted, turned to fly, Meek, who was riding an old mule that had to be beaten over the head to make it go, seeing that he was going to be left behind, called out lustily, "Hold on, boys! There's not many of them. Let's stop and fight 'em," at the same time pounding the mule over the head, but without effect. The Indians saw the predicament, and ran up to seize the mule by the bridle, but the moment the mule got wind of the savages, away he went, racing like a thoroughbred, jumping impediments, and running right over a ravine, which was fortunately filled with snow. This movement brought Meek out ahead.

The other men then began to call out to Meek to stop and fight. "Run for your lives, boys," roared Meek back at them. "There's ten thousand of them, they'll kill every one of you!"

The mule had got his head, and there was no more stopping him than there had been starting him. On he went in the direction of the Yellowstone, while the others made for Clarke's Fork. On arriving at the former river, Meek found that some of the packhorses had followed him, and others the rest of the party.

This had divided the Indians, three or four of whom were on his trail. Springing off his mule, he threw his blankets down on the ice, and by moving them alternately, soon crossed the mule over to the opposite side, just in time to avoid a bullet that came whistling after him. As the Indians could not follow, he pursued his way to camp in safety, arriving late that evening. The main party were already in and expecting him. Soon after, the buffalo hunters returned to the big camp, minus some packhorses, but with a good story to tell, at the expense of Meek, and which he enjoys telling of himself to this day.

# 12

## 1835

Owing to the high rate of pay which Meek was now able to command, he began to think of imitating the example of that distinguished order, the free trappers, to which he now belonged, and setting up a lodge to himself as a family man. The writer of this veracious history has never been able to obtain a full and particular account of our hero's earliest love adventures. This is a subject on which, in common with most mountain men, he observes a becoming reticence. But of one thing we feel quite well assured: that from the time when the young Shoshone beauty assisted in the rescue of himself and Sublette from the execution of the death sentence at the hands of her people, Meek had always cherished a rather more than friendly regard for the *Mountain Lamb*.

But Sublette, with wealth and power, and the privileges of a

Booshway, had hastened to secure her for himself and Meek had to look and long from afar off, until, in the year of which we are writing, Milton Sublette was forced to leave the mountains and repair to an eastern city for surgical aid, having received a very troublesome wound in the leg, which was only cured at last by amputation.

Whether it was the act of a gay Lothario, or whether the law of divorce is even more easy in the mountains than in Indiana, we have always judiciously refrained from inquiring, but this we do know, upon the word of Meek himself, no sooner was Milton's back turned, than his friend so insinuated himself into the good graces of his *Isabel*, as Sublette was wont to name the lovely Umentucken, that she consented to join her fortunes to those of the handsome young trapper without even the ceremony of serving a notice on her former lord. As their season of bliss only extended over one brief year, this chapter shall be entirely devoted to recording such facts as have been imparted to us concerning this free trapper's wife.

"She was the most beautiful Indian woman I ever saw," says Meek. "And when she was mounted on her dapple gray horse, which cost me three hundred dollars, she made a fine show. She wore a skirt of beautiful blue broadcloth, and a bodice and leggins' of scarlet cloth, of the very finest make. Her hair was braided and fell over her shoulders, a scarlet silk handkerchief, tied on hood fashion, covered her head, and the finest embroidered moccasins on her feet. She rode like all the Indian women, astride, and carried on one side of the saddle the tomahawk for war, and on the other the pipe of peace.

"The name of her horse was *All Fours.* His accoutrements

were as fine as his rider's. The saddle, crupper, and bust girths cost one hundred and fifty dollars, the bridle fifty dollars, and the musk-a-moots fifty dollars more. All these articles were ornamented with fine-cut glass beads, porcupine quills, and hawk's bells, that tinkled at every step. Her blankets were of scarlet and blue, and of the finest quality. Such was the outfit of the trapper's wife, *Umentucken, Tukutey Undenwatsy*, the Lamb of the Mountains."

Although Umentucken was beautiful, and had a name signifying gentleness, she was not without a will and a spirit of her own, when the occasion demanded it. While the camp was on the Yellowstone River, in the summer of 1835, a party of women left it to go in search of berries, which were often dried and stored for winter use by the Indian women. Umentucken accompanied this party, which was attacked by a band of Blackfeet, some of the squaws being taken prisoners. But Umentucken saved herself by flight, and by swimming the Yellowstone while a hundred guns were leveled on her, the bullets whistling about her ears.

At another time she distinguished herself in camp by a quarrel with one of the trappers, in which she came off with flying colors. The trapper was a big, bullying Irishman named O'Fallen, who had purchased two prisoners from the Snake Indians, to be kept in a state of slavery, after the manner of the savages. The prisoners were Utes, or Utahs, who soon contrived to escape. O'Fallen, imagining that Umentucken had liberated them, threatened to whip her, and armed himself with a horsewhip for that purpose. On hearing of these threats, Umentucken repaired to her lodge, and also armed herself, but with a pistol.

When O'Fallen approached, the whole camp looking on to see the event, Umentucken slipped out at the back of the lodge and coming around, confronted him before he could enter.

"Coward!" she cried. "You would whip the wife of Meek. He is not here to defend me, not here to kill you. But I shall do that for myself," and with that, she presented the pistol to his head. O'Fallen taken by surprise, and having every reason to believe she would keep her word, and kill him on the spot, was obliged not only to apologize, but to beg to have his life spared. This Umentucken consented to do on condition of his sufficiently humbling himself, which he did in a very shame-faced manner, and a shout then went up from the whole camp, "Hurrah for the Mountain Lamb!" For nothing more delights a mountaineer than a show of pluck, especially in an unlooked-for quarter.

The Indian wives of the trappers were often in great peril, as well as their lords. Whenever it was convenient, they followed them on their long marches through dangerous countries. But if the trapper was only going out for a few days, or if the march before him was more than usually dangerous, the wife remained with the main camp.

During this year of which we are writing, a considerable party had been out on Powder River hunting buffalo, taking their wives along with them. When on the return, just before reaching camp, Umentucken was missed from the cavalcade. She had fallen behind, and been taken prisoner by a party of twelve Crow Indians. As soon as she was missed, a volunteer party mounted their buffalo horses in such haste that they waited not for saddle or bridle, but snatched only a halter, and started back in pursuit. They had not run a very long distance when they discovered poor Umentucken in the midst of her jubilant captors, who were delighting their eyes with gazing at her fine feathers, and promising themselves very soon to pluck the gay bird, and appropriate her trinkets to their own use.

Their delight was premature. Swift on their heels came an avenging, as well as a saving spirit. Meek, at the head of his six comrades, no sooner espied the drooping form of the Lamb, than he urged his horse to the top of its speed. The horse was a spirited creature, that seeing something wrong in all these hasty maneuvers, took fright and adding terror to good will, ran with the speed of madness right in among the startled Crows, who doubtless regarded as a great *medicine* so fearless a warrior. It was now too late to be prudent, and Meek began the battle by yelling and firing, taking care to hit his Indian. The other trappers, emulating the bold example of their leader, dashed into the melee and a chance-medley fight was carried on, in which Umentucken escaped, and another Crow bit the dust. Finding that they were getting the worst of the fight, the Indians at length took to flight, and the trappers returned to camp rejoicing, and complimenting Meek on his gallantry in attacking the Crows single-handed.

"I took their compliments quite naturally," says Meek, "nor did I think it war worthwhile to explain to them that I couldn't hold my horse."

The Indians are lordly and tyrannical in their treatment of women, thinking it no shame to beat them cruelly, even taking the liberty of striking other women than those belonging to their own families. While the camp was traveling through the Crow country in the spring of 1836, a party of that nation paid a visit to Bridger, bringing skins to trade for blankets and ammunition.

The bargaining went on quite pleasantly for some time, but one of the braves who was promenading about camp inspecting whatever came in his way, chanced to strike Umentucken with a whip he carried in his hand, by way of displaying his superiority to squaws in general, and trappers' wives in particular. It was an unlucky blow for the brave, for in another instant, he rolled on the ground, shot dead by a bullet from Meek's gun.

At this rash act, the camp was in confusion. Yells from the Crows, who took the act as a signal for war, and hasty questions, and cries of command, also arming and shooting. It was some time before the case could be explained or understood. The Crows had two or three of their party shot. The whites also lost a man. After the unpremeditated fight was over, and the Crows departed not thoroughly satisfied with the explanation, Bridger went round to Meek's lodge.

"Well, you raised a hell of a row in camp," said the commander, rolling out his deep bass voice in the slow, monotonous tones which mountain men very quickly acquire from the Indians.

"Very sorry, Bridger, but couldn't help it. No devil of an Indian shall strike Meek's wife."

"But you got a man killed."

"Sorry for the man, couldn't help it, though, Bridger."

And in truth, it was too late to mend the matter. Fearing, however, that the Crows would attempt to avenge themselves for the losses they had sustained, Bridger hurried his camp forward, and got out of their neighborhood as quickly as possible.

So much for the female element in the camp of the Rocky Mountain trapper. Woman, it is said, has held the apple of discord, from mother Eve to Umentucken, and in consonance with this theory, Bridger, doubtless, considered the latter as the primal cause of the unfortunate *row in camp,* rather than the brutality of the Crow, or the imprudence of Meek.

But Umentucken's career was nearly run. In the following summer, she met her death by a Bannack arrow, dying like a warrior, although living she was only a woman.

# 13

## 1835

The rendezvous of the Rocky Mountain Company seldom took place without combining with its many wild elements, some other more civilized and refined. Artists, botanists, travelers, and hunters, from the busy world outside the wilderness, frequently claimed the companionship, if not the hospitality of the fur companies, in their wanderings over prairies and among mountains. Up to the year 1835, these visitors had been of the classes just named, men traveling either for the love of adventure, to prosecute discoveries in science, or to add to art the treasure of new scenes and subjects.

But in this year,[1] there appeared at rendezvous two gentlemen, who had accompanied the St. Louis Company in its

1. This rendezvous of 1835 was held on the upper Green River, the exact location unknown.

outward trip to the mountains, whose object was not the procurement of pleasure, or the improvement of science. They had come to found missions among the Indians—Rev. Samuel Parker and Rev. Dr. Marcus Whitman, the first a scholarly and fastidious man, and the other possessing all the boldness, energy, and contempt of fastidiousness, which would have made him as good a mountain leader, as he was an energetic servant of the American Board of Foreign Missions.

The cause which had brought these gentlemen to the wilderness was a little incident connected with the fur trade. Four Flathead Indians, in the year 1832, having heard enough of the Christian religion, from the few devout men connected with the fur companies, to desire to know more, performed a winter journey to St. Louis, and there made inquiry about the white man's religion. This incident, which to anyone acquainted with Indian character, would appear a very natural one, when it became known to Christian churches in the United States, excited a very lively interest, and seemed to call upon them like a voice out of heaven, to fly to the rescue of perishing heathen souls. The Methodist Church was the first to respond. When Wyeth returned to the mountains in 1834, four missionaries accompanied him, destined for the valley of the Willamette River in Oregon. In the following year, the Presbyterian Church sent out its agents, the two gentlemen above mentioned, one of whom, Dr. Whitman, subsequently located near Fort Walla-Walla.

The account given by Capt. Bonneville of the Flatheads and Nez Percé , as he found them in 1832, before missionary labor had been among them, throws some light on the incident of the journey to St. Louis, which so touched the Christian heart in the

United States. After relating his surprise at finding that the Nez Percé observed certain sacred days, he continues, "A few days afterward, four of them signified that they were about to hunt. 'What!' exclaimed the captain, 'without guns or arrows, and with only one old spear? What do you expect to kill?' They smiled among themselves, but made no answer. Preparatory to the chase, they performed some religious rights, and offered up to the Great Spirit a few short prayers for safety and success, then having received the blessing of their wives, they leaped upon their horses and departed, leaving the whole party of Christian spectators amazed and rebuked by this lesson of faith and dependence on a supreme and benevolent being. Accustomed as I had heretofore been to find the wretched Indian reveling in blood, and stained by every vice which can degrade human nature, I could scarcely realize the scene which I had witnessed. Wonder at such unaffected tenderness and piety, where it was least to have been sought, contended in all our bosoms with shame and confusion, at receiving such pure and wholesome instructions from creatures so far below us in all the arts and comforts of life.

"Simply to call these people religious," continued Bonneville, "would convey but a faint idea of the deep hue of piety and devotion which pervades their whole conduct. Their honesty is immaculate, and their purity of purpose, and their observance of the rites of their religion, are most uniform and remarkable. They are certainly more like a nation of saints than a horde of savages."

This was a very enthusiastic view to take of the Nez Percé character, which appeared all the brighter to the captain, by contrast with the savage life which he had witnessed in other places, and even by contrast with the conduct of the white trappers. But the Nez Percé and Flatheads were, intellectually and morally, an exception to all the Indian tribes west of the Missouri River. Lewis and Clarke found them different from any others, the fur traders and the missionaries found them different and they remain at this day an honorable example for probity and piety to both savage and civilized peoples.

To account for this superiority is indeed difficult. The only clue to the cause is in the following statement of Bonneville's. "It would appear," he says, "that they had imbibed some notions of the Christian faith from Catholic missionaries and traders who had been among them. They even had a rude calendar of the fasts and festivals of the Romish Church, and some traces of its ceremonials. These have become blended with their own wild rites, and present a strange medley, civilized and barbarous."[2]

Finding that these people among whom he was thrown exhibited such remarkable traits of character, Captain Bonneville exerted himself to make them acquainted with the history and spirit of Christianity. To these explanations, they listened with great eagerness. "Many a time," he says, "was my little lodge thronged, or rather piled with hearers, for they lay on the ground, one leaning over the other, until there was no further room, all listening with greedy ears to the wonders which the Great Spirit had revealed to the white man. No other subject gave them half the satisfaction, or commanded half the attention, and but few scenes of my life remain so freshly on my

---

2. This claim that the Indian interest in Christianity is purely material is probably one of the bold bits that got the author in trouble with the missionary faction in Oregon.

memory, or are so pleasurably recalled to my contemplation, as these hours of intercourse with a distant and benighted race in the midst of the desert."

It was the interest awakened by these discourses of Captain Bonneville, and possibly by Smith, and other traders who happened to fall in with the Nez Perces and Flatheads, that stimulated those four Flatheads to undertake the journey to St. Louis in search of information, and this it was which resulted in the establishment of missions, both in western Oregon, and among the tribes inhabiting the country between the two great branches of the Columbia.

The trait of Indian character which Bonneville, in his pleased surprise at the apparent piety of the Nez Perces and Flatheads, failed to observe, and which the missionaries themselves for a long time remained oblivious to, was the material nature of their religious views. The Indian judges of all things by the material results. If he is possessed of a good natural intelligence and powers of observation, he soon discovers that the God of the Indian is but a feeble deity, for does he not permit the Indian to be defeated in war, to starve and to freeze? Do not the Indian medicine men often fail to save life, to win battles, to curse their enemies? The Indian's God, he argues, must be a good deal of a humbug. He sees the white men faring much better. They have guns, ammunition, blankets, knives, everything in plenty, and they are successful in war and are skillful in a thousand things the Indian knows nothing of. To be so blessed implies a very wise and powerful deity. To gain all these things, they are eager to learn about the white man's God and are willing to do whatever is necessary to please and propitiate Him. Hence, their attentiveness to the white man's discourse about his religion.

Naturally enough, they were struck with wonder at the doctrine of peace and good will, a doctrine so different from the law of blood by which the Indian, in his natural state, lives. Yet if it is good for the white men, it must be good for him, for at all events, he is anxious to try it.

That is the course of reasoning by which an Indian is led to inquire into Christianity. It is a desire to better his physical, rather than his spiritual condition, for of the latter, he has but a very faint conception. He was accustomed to desire a material heaven, such a world beyond the grave, as he could only imagine from his earthly experience. Heaven was happiness, and happiness was plenty, therefore the most a good Indian could desire was to go where there should forevermore be plenty.

Such was the Indian's view of religion, and it could be no other. Until the wants of the body have been supplied by civilization, the wants of the soul do not develop themselves, and until then, the savage is not prepared to understand Christianity. This is the law of nature and of God. Primeval man was a savage, and it was little by little, through thousands of years, that Christ was revealed. Every child born, even now, is a savage, and has to be taught civilization year after year, until he arrives at the possibility of comprehending spiritual religion. So every full-grown barbarian is a child in moral development, and to expect him to comprehend those mysteries over which the world has agonized for centuries, is to commit the gravest error. Into this error fell all the missionaries who came to the wilds that lay beyond the Rocky Mountains. They undertook to teach religion first, and more simple matters afterward—building their edifice like the Irishman's chimney, by holding up the top brick, and putting the others under it. Failure was the result of such a process, as the record of the Oregon Missions sufficiently proves.

The reader will pardon this digression—made necessary by the part which one of the gentlemen present at this year's rendezvous, was destined to take in the history which we are writing. Shortly after the arrival of Messrs. Parker and Whitman, rendezvous broke up. A party, to which Meek was attached, moved in the direction of the Snake River headwaters, the missionaries accompanying them, and after making two camps, came on Saturday eve to Jackson's Little Hole, a small mountain valley near the larger one commonly known as Jackson's Hole.

On the following day religious services were held in the Rocky Mountain camp. A scene more unusual could hardly have transpired than that of a company of trappers listening to the preaching of the Word of God. Very little pious reverence marked the countenances of that wild and motley congregation. Curiosity, incredulity, sarcasm, or a mocking levity, were more plainly perceptible in the expression of the men's faces, than either devotion or the longing expectancy of men habitually deprived of what they once highly valued. The Indians alone showed by their eager listening that they desired to become acquainted with the mystery of the *Unknown God.*

The Reverend Samuel Parker preached, and the men were as politely attentive as it was in their reckless natures to be, until, in the midst of the discourse, a band of buffalo appeared in the valley, when the congregation incontinently broke up, without staying for a benediction, and every man-made haste after his horse, gun, and rope, leaving Mr. Parker to discourse to vacant ground.

The run was both exciting and successful. About twenty fine buffaloes were killed, and the choice pieces brought to camp, cooked and eaten, amid the merriment, mixed with something coarser, of the hunters. On this noisy rejoicing, Mr. Parker looked with a sober aspect, and following the dictates of his religious feeling, he rebuked the sabbath-breakers quite severely. Better for his influence among the men, if he had not done so, or had not eaten so heartily of the tender-loin afterward, a circumstance which his irreverent critics did not fail to remark, to his prejudice, and upon the principle that the *partaker is as bad as the thief*, they set down his lecture on sabbath-breaking as nothing better than pious humbug.

Dr. Marcus Whitman was another style of man. Whatever he thought of the wild ways of the mountain men he discreetly kept to himself, preferring to teach by example rather than precept, and showing no fastidious contempt for any sort of rough duty he might be called upon to perform. So aptly indeed had he turned his hand to all manner of camp service on the journey to the mountains, that this abrogation of clerical dignity had become a source of solicitude, not to say disapproval and displeasure on the part of his colleague, and it was agreed between them that the Doctor should return to the States with the St. Louis Company, to procure recruits for the promising field of labor which they saw before them, while Mr. Parker continued his journey to the Columbia to decide upon the location of the missionary stations. The difference in character of the two men was clearly illustrated by the results of this understanding. Parker went to Vancouver, where he was hospitably entertained, and where he could inquire into the workings of the missionary system as pursued by the Methodist missionaries.

His investigations not proving the labor to his taste, he sailed the following summer for the Sandwich Islands, and thence to New York, leaving only a brief note for Dr. Whitman, when he, with indefatigable exertions, arrived that season among the Nez Perces with a missionary company, eager for the work which they hoped to make as great as they believed it to be good.

# 14

From the mountains about the headwaters of the Snake River, Meek returned with Bridger's brigade to the Yellowstone country, where he fell into the hands of the Crows. The story as he relates it, is as follows:

"I war trapping on the Rocky Fork of the Yellowstone. I had been out from camp five days and war solitary and alone, when I war discovered by a war party of Crows. They had the prairie, and I war forced to run for the creek bottom, but the beaver had throwed the water out and made dams, so that my mule mired down. While I war struggling in the marsh, the Indians came after me, with tremendous yells, firing a random shot now and then, as they closed in on me.

"When they war within about two rods of me, I brought old *Sally*, that is my gun, to my face, ready to fire, and then die—for I knew it war death this time, unless Providence interfered to save me: and I didn't think Providence would do it. But the head chief, when he saw the warlike looks of *Sally*, called out to me to put down my gun, and I should live.

"Well, I liked to live, being then in the prime of life, and though it hurt me powerful, I resolved to part with *Sally*. I laid her down. As I did so, the chief picked her up, and one of the braves sprang at me with a spear, and would have run me through, but the chief knocked him down with the butt of my gun. Then they led me forth to the high plain on the south side of the stream. There, they called a halt, and I was given in charge of three women, while the warriors formed a ring to smoke and consult. This gave me an opportunity to count them: they numbered one hundred and eighty-seven men, nine boys, and three women.

"After a smoke of three long hours, the chief, who war named *The Bold*, called me in the ring, and said:

"'I have known the whites for a long time, and I know them to be great liars, deserving death, but if *you* will tell the truth, you shall live.'

"Then I thought to myself, they will fetch the truth out of me, if thar is any in me. But his Highness continued:

"'Tell me whar are the whites you belong to and what is your captain's name.'

"I said 'Bridger is my captain's name, or in the Crow tongue, *Casapy*, the *Blanket chief*.' At this answer, the chief seemed lost in thought. At last, he asked me—

"'How many men has he?'

"I thought about telling the truth and living, but I said 'forty,' which war a tremendous lie, for thar war two hundred and forty. At this answer, The Bold laughed:

"'We will make them poor,' said he, 'and you shall live, but they shall die.'

"I thought to myself, *hardly*, but I said nothing. He then asked me whar I war to meet the camp, and I told him—and then how many days before the camp would be thar, which I answered truly, for I wanted them to find the camp.

"It war now late in the afternoon, and thar war a great bustle, getting ready for the march to meet Bridger. Two big Indians mounted my mule, but the women made me pack moccasins. The spies started first, and after a while, the main party. Seventy warriors traveled ahead of me. I war placed with the women and boys, and after us the balance of the braves. As we traveled along, the women would prod me with sticks, and laugh, and say *Masta Sheela*—which means white man. 'Masta sheela very poor now.' The fair sex war very much amused.

"We traveled that way till midnight, the two big bucks riding my mule, and I packing moccasins. Then we camped, the Indians in a ring, with me in the center, to keep me safe. I didn't sleep very well that night. I'd a heap rather been in some other place.

"The next morning, we started on in the same order as before, and the squaws making fun of me all day, but I kept mighty quiet. When we stopped to cook that evening, I war set to work, and war head cook, and head waiter too. The third and the fourth day it war the same. I felt pretty bad when we struck camp oh the last day, for I knew we must be coming near to Bridger, and that if anything should go wrong, my life would pay the forfeit.

"On the afternoon of the fourth day, the spies, who war in advance, looking out from a high hill, made a sign to the main party. In a moment, all sat down. Directly, they got another sign, and then they got up and moved on. I war as well up in Indian signs as they war, and I knew they had discovered white men. What war worse, I knew they would soon discover that I had been lying to them. All I had to do then war to trust to luck. Soon we came to the top of the hill, which overlooked the Yellowstone, from which I could see the plains below extending as far as the eye could reach, and about three miles off, the camp of my friends. My heart beat double quick about that time, and I once in a while put my hand to my head, to feel if my scalp war thar.

"While I war watching our camp, I discovered that the horse guard had seen us for I knew the sign he would make if he discovered Indians. I thought the camp a splendid sight that evening. It made a powerful show to me, who did not expect ever to see it after that day. And it *war* a fine sight anyhow, from the hill whar I stood. About two hundred and fifty men, and women and children in great numbers, and about a thousand horses and mules. Then the beautiful plain, and the sinking sun, and the herds of buffalo that could not be numbered, and the cedar hills, covered with elk—I never saw so fine a sight as all that looked to me then!

"When I turned my eyes on that savage Crow band, and saw the chief standing with his hand on his mouth, lost in amazement, and beheld the warriors' tomahawks and spears glittering in the sun, my heart war very little. Directly, the chief turned to me with a horrible scowl. Said he:

"'I promised that you should live if you told the truth, but you have told me a great lie.'

"Then the warriors gathered around, with their tomahawks

in their hands, but I war showing off very brave, and kept my eyes fixed on the horse guard who war approaching the hill to drive in the horses. This drew the attention of the chief, and the warriors too. Seeing that the guard war within about two hundred yards of us, the chief turned to me and ordered me to tell him to come up. I pretended to do what he said, but instead of that I howled out to him to stay off, or he would be killed, and to tell Bridger to try to treat with them, and get me away.

"As quick as he could he ran to camp, and in a few minutes, Bridger appeared, on his large white horse. He came up to within three hundred yards of us, and called out to me, asking who the Indians war. I answered, 'Crows.' He then told me to say to the chief he wished him to send one of his sub-chiefs to smoke with him.

"All this time my heart beat terribly hard. I don't know now why they didn't kill me at once, but the head chief seemed overcome with surprise. When I repeated to him what Bridger said, he reflected a moment, and then ordered the second chief, called Little-Gun, to go and smoke with Bridger. But they kept on preparing for war, getting on their paint and feathers, arranging their scalp locks, selecting their arrows, and getting their ammunition ready.

"While this war going on, Little-Gun had approached to within about a hundred yards of Bridger, when, according to the Crow laws of war, each war forced to strip himself, and proceed the remaining distance in a state of nudity, and kiss and embrace. While this interesting ceremony war being performed, five of Bridger's men had followed him, keeping in a ravine until they got within shooting distance, when they showed themselves, and cut off the return of Little-Gun, thus making a prisoner of him.

"If you think my heart did not jump up when I saw that, you think wrong. I knew it war kill or cure, now. Every Indian snatched a weapon, and fierce threats war howled against me. But all at once, about a hundred of our trappers appeared on the scene. At the same time Bridger called to me, to tell me to propose to the chief to exchange me for Little-Gun. I explained to The Bold what Bridger wanted to do, and he sullenly consented, for, he said, he could not afford to give a chief for one white dog's scalp. I war then allowed to go toward my camp, and Little-Gun toward his, and the rescue I hardly hoped for war accomplished.

"In the evening, the chief, with forty of his braves, visited Bridger and made a treaty of three months. They said they war formerly at war with the whites, but that they desired to be friendly with them now, so that together they might fight the Blackfeet, who war everybody's enemies. As for me, they returned me my mule, gun, and beaver packs, and said my name should be *Shiam Shaspusia*, for I could out-lie the Crows."

In December, Bridger's command went into winter quarters in the bend of the Yellowstone.[1] Buffalo, elk, and bear were in great abundance, all that fall and winter. Before they went to camp, Meek, Kit Carson, Hawkins, and Doughty were trapping

1. Osborne Russell says the Bridger brigade made winter camp on Blackfoot Creek in 1835-36, a tributary of Snake River, and that the men had little to eat that winter. Most authorities agree with Russell. Someone, probably Meek, has his years confused. Joe's description of this camp at a big bend of the Yellowstone, with plenty of meat, seems to fit the next winter. Except for events at rendezvous, in fact, Joe's dates seem a year behind from Chapter 11 to here.

together on the Yellowstone, about sixty miles below. They had made their temporary camp in the ruins of an old fort, the walls of which were about six feet high. One evening, after coming in from setting their traps, they discovered three large grizzly bears in the river bottom, not more than half a mile off, and Hawkins went out to shoot one. He was successful in killing one at the first shot, when the other two, taking fright, ran toward the fort. As they came near enough to show that they were likely to invade camp, Meek and Carson, not caring to have a bear fight, clambered up a cottonwood tree close by, at the same time advising Doughty to do the same. But Doughty was tired, and lazy besides, and concluded to take his chances where he was, so he rolled himself in his blanket and laid quite still. The bears, on making the fort, reared up on their hind legs and looked in as if meditating, taking it for a defense.

The sight of Doughty lying rolled in his blanket, and the monster grizzlys inspecting the fort, caused the two trappers who were safely perched in the cottonwood to make merry at Doughty's expense, saying all the mirth-provoking things they could, and then advising him not to laugh, for fear the bears should seize him. Poor Doughty, agonizing between suppressed laughter and growing fear, contrived to lie still however, while the bears gazed upward at the speakers in wonder, and alternately at the suspicious-looking bundle inside the fort. Not being able to make out the meaning of either, they gave at last a grunt of dissatisfaction, and ran off into a thicket to consult over these strange appearances, leaving the trappers to enjoy the incident as a very good joke. For a long time after, Doughty was reminded how close to the ground he laid, when the grizzlys paid their compliments to him. Such were the everyday incidents from which the mountain men contrived to derive their rude jests, and laughter-provoking reminiscences.

A few days after this incident, while the same party were trapping a few miles farther down the river, on their way to camp, they fell in with some Delaware Indians, who said they had discovered signs of Blackfeet, and wanted to borrow some horses to decoy them. To this, the trappers very willingly agreed, and they were furnished with two horses. The Delawares then went to the spot where signs had been discovered, and tying the horses, laid flat down on the ground near them, concealed by the grass or willows. They had not long to wait before a Blackfoot was seen stealthily advancing through the thicket, confident in the belief that he should gain a couple of horses while their supposed owners were busy with their traps.

But just as he laid his hand on the bridle of the first one, crack went the rifles of the Delawares, and there was one less Blackfoot thief on the scent after trappers. As soon as they could, after this, the party mounted and rode to camp, not stopping by the way, lest the main body of Blackfeet should discover the deed and seek for vengeance. Truly indeed, was the Blackfoot the Ishmael of the wilderness, whose hand was against every man, and every man's hand against him.

The Rocky Mountain Company passed the first part of the winter in peace and plenty in the Yellowstone camp, unannoyed either by enemies or rivals. Hunting buffalo, feeding their horses, playing games, and telling stories, occupied the entire leisure of these months of repose. Not only did the mountain men recount their own adventures, but when these were exhausted, those whose memories served them rehearsed the tales they had read in their youth. Robinson Crusoe and the

Arabian Nights Entertainment, were read over again by the light of memory, and even Bunyan's Pilgrim's Progress was made to recite like a sensation novel, and was quite as well enjoyed.

---

## 1836

In January, however, this repose was broken in upon by a visit from the Blackfeet. As their visitations were never of a friendly character, so then they were not bent upon Pacific rites and ceremonies, such as all the rest of the world find pleasure in, but came in full battle array to try their fortunes in war against the big camp of the whites. They had evidently made great preparation. Their warriors numbered eleven hundred, got up in the top of the Blackfoot fashions, and armed with all manner of savage and some civilized weapons. But Bridger was prepared for them, although their numbers were so overwhelming. He built a fort, had the animals corralled, and put himself on the defensive in a prompt and thorough manner. This made the Blackfeet cautious, and they, too, built forts of cottonwood in the shape of lodges, ten men to each fort, and carried on a skirmishing fight for two days, when finding there was nothing to be gained, they departed, neither side having sustained much loss, with the whites losing only two men by this grand Blackfoot army.

Soon after this attack Bridger broke camp, and traveled up the Yellowstone, through the Crow country. It was while on this march that Umentucken was struck by a Crow, and Meek put the whole camp in peril, by shooting him. They passed on to the

Big Horn and Little Horn rivers, down through the Wind River Valley and through the South Pass to Green River.

While in that country, there occurred the fight with the Bannacks in which Umentucken was killed. A small party of Nez Perces had lost their horses by the thieving of the Bannacks. They came into camp and complained to the whites, who promised them their protection, should they be able to recover their horses. Accordingly, the Nez Perces started after the thieves, and by dogging their camp, succeeded in re-capturing their horses and getting back to Bridger's camp with them. In order to divert the vengeance of the Bannacks from themselves, they presented their horses to the whites, and a very fine one to Bridger.

All went well for a time. The Bannacks went on their way to hunt buffalo, but they treasured up their wrath against the supposed white thieves who had stolen the horses which they had come by so honestly. On their return from the hunt, having learned by spies that the horses were in the camp of the whites, they prepared for war. Early one morning, they made their appearance mounted and armed, and making a dash at the camp, rode through it with the usual yells and frantic gestures. The attack was entirely unexpected. Bridger stood in front of his lodge, holding his horse by a lasso, and the head chief rode over it, jerking it out of his hand. At this unprecedented insult to his master, a negro named Jim, cook to the Booshways, seized a rifle and shot the chief dead. At the same time, an arrow shot at random struck Umentucken in the breast, and the joys and sorrows of the Mountain Lamb were over forevermore.

The killing of a head chief always throws an Indian war party into confusion, and negro Jim was greatly elated at this signal feat of his. The trappers, who were as much surprised at the suddenness of the assault as it is in the mountain man's nature to be, quickly recovered themselves. In a few moments the men were mounted and in motion, and the disordered Bannacks were obliged to fly toward their village, Bridger's company pursuing them.

All the rest of that day, the trappers fought the Bannacks, driving them out of their village and plundering it, and forcing them to take refuge on an island in the river. Even there they were not safe, the guns of the mountain men picking them off from their stations on the river banks. Umentucken was well avenged that day.

All night, the Indians remained on the island, where sounds of wailing were heard continually, and when morning came one of their old women appeared bearing the pipe of peace. "You have killed all our warriors," she said. "Do you now want to kill the women? If you wish to smoke with women, I have the pipe."

Not caring either to fight or to smoke with so feeble a representative of the Bannacks, the trappers withdrew. But it was the last war party that nation ever sent against the mountain men, though in later times, they have by their atrocities, avenged the losses of that day.

While awaiting, in the Green River valley, the arrival of the St. Louis Company, the Rocky Mountain and North American companies united, after which Captain Sublette and his brother returned no more to the mountains. The new firm was known only as the American Fur Company, the other having dropped its title altogether. The object of their consolidation was by combining their capital and experience to strengthen their hands against the Hudson's Bay Company, which now had an establishment at Fort Hall, on the Snake River. By this new arrangement, Bridger and Fontenelle commanded, and Dripps was to be the traveling partner who was to go to St. Louis for goods.

After the conclusion of this agreement, Dripps, with the restlessness of the true mountain man, decided to set out, with a small party of equally restless trappers, always eager to volunteer for any undertaking promising either danger or diversion, to look for the St. Louis Company which was presumed to be somewhere between the Black Hills and Green River. According to this determination Dripps, Meek, Carson, Newell, a Flathead chief named Victor, and one or two others, set out on the search for the expected company.

It happened, however, that a war party of a hundred Crows were out on the trail before them, looking perhaps for the same party, and the trappers had not made more than one or two camps before they discovered signs which satisfied them of the neighborhood of an enemy. At their next camp on the Sandy, Meek, and Carson, with the caution and vigilance peculiar to them, kept their saddles on their horses, and the horses tied to themselves by a long rope, so that on the least unusual motion of the animals they should be readily informed of the disturbance. Their precaution was not lost. Just after midnight had given place to the first faint kindling of dawn, their ears were stunned by the simultaneous discharge of a hundred guns, and the usual furious din of the war-whoop and yell. A stampede immediately took place of all the horses excepting those of Meek and Carson. "Every man for himself and God for us all," is the motto of the mountain man in case of an Indian attack, nor did our trappers forget it on this occasion. Quickly mounting, they put their horses to their speed, which was not checked until they had left the Sandy far behind them. Continuing on in the direction of the proposed meeting with the St. Louis Company, they made their first camp on the Sweetwater, where they fell in with Victor, the Flathead chief, who had made his way on foot to this place. One or two others came into camp that night, and the following day, this portion of the party traveled on in company until within about five miles of Independence Rock, when they were once more charged on by the Indians, who surrounded them in such a manner that they were obliged to turn back to escape.

Again, Meek and Carson made off, leaving their dismounted comrades to their own best devices. Finding that with so many Indians on the trail, and only two horses, there was little hope of being able to accomplish their journey, these two lucky ones made all haste back to camp. On Horse Creek, a few hours travel from rendezvous, they came up with Newell, who, after losing his horse, had fled in the direction of the main camp, but becoming bewildered, had been roaming about until he was quite tired out, and on the point of giving up. But as if the creek where he was found meant to justify itself for having so inharmonious a name, one of their own horses, which had escaped from the Crows was found quietly grazing on its banks, and the worn-out fugitive at once remounted. Strange as it may appear, not one of the party was killed, the others returning to camp two days later than Meek and Carson, the worse for their expedition only by the loss of their horses, and rather an unusually fatigued and forlorn aspect.

# 15

### 1836

While the resident partners of the consolidated company waited at the rendezvous for the arrival of the supply trains from St. Louis, word came by a messenger sent forward, that the American Company under Fitzpatrick, had reached Independence Rock and was pressing forward. The messenger also brought the intelligence that two other parties were traveling in company with the fur company—that of Captain Stuart, who had been to New Orleans to winter, and that of Dr. Whitman, one of the missionaries who had visited the mountains the year previous. In this latter party, it was asserted, there were two white ladies.

This exhilarating news immediately inspired some of the trappers, foremost among whom was Meek, with a desire to be the first to meet and greet the on-coming caravan, and especially to salute the two white women who were bold enough to invade a mountain camp. In a very short time Meek, with half-a-dozen comrades, and ten or a dozen Nez Perces, were mounted and away, on their self-imposed errand of welcome, also the trappers because they were *spoiling* for a fresh excitement, and the Nez Perces because the missionaries were bringing them information concerning the powerful and beneficent deity of the white men. These latter also were charged with a letter to Dr. Whitman from his former associate, Mr. Parker.

On the Sweetwater about two days' travel from camp the caravan of the advancing company was discovered, and the trappers prepared to give them a characteristic greeting. To prevent mistakes in recognizing them, a white flag was hoisted on one of their guns, and the word was given to start. Then over the brow of a hill they made their appearance, riding with that mad speed only an Indian or a trapper can ride, yelling, whooping, dashing forward with frantic and threatening gestures. Their dress, noises, and motions, all so completely savage that the white men could not have been distinguished from the red.

The first effect of their onset was what they probably intended. The uninitiated travelers, including the missionaries, believing they were about to be attacked by Indians, prepared for defense, nor could be persuaded that the preparation was unnecessary until the guide pointed out to them the white flag in advance. At the assurance that the flag betokened friends, apprehension was changed to curiosity and intense interest. Every movement of the wild brigade became fascinating. On they came, riding faster and faster, yelling louder and louder, and gesticulating more and more madly, until, as they met and passed the caravan, they discharged their guns in one volley over the heads of the company, as a last finishing *feu de joie,* and suddenly wheeling rode back to the front as wildly as they had come. Nor could this first brief display content the crazy cavalcade. After reaching the front, they rode back and forth, and around and around the caravan, which had returned their salute, showing off their feats of horsemanship, and the knowing tricks of their horses together, and hardly stopping to exchange questions and answers, but seeming really intoxicated with delight at the meeting. What strange emotions filled the breasts of the lady missionaries when they beheld among whom their lot was cast may now be faintly outlined by a vivid imagination but have never been, perhaps never could be put into words.

The caravan on leaving the settlements, had consisted of nineteen laden carts, each drawn by two mules driven tandem, and one light wagon, belonging to the American Company, two wagons with two mules to each, belonging to Capt. Stuart, and one light two-horse wagon, and one four-horse freight wagon, belonging to the missionaries. However, all the wagons had been left behind at Fort Laramie, except those of the missionaries, and one of Capt. Stuart's, so that the three that remained in the train when it reached the Sweetwater were alone in the enjoyment of the Nez Percé's curiosity concerning them. This was a curiosity which they divided between them and the domesticated cows and calves belonging to the missionaries, another proof, as they considered it, of the superior power of the white man's God, who could give to the whites the ability to tame wild animals to their uses.

But it was toward the two missionary ladies, Mrs. Whitman and Mrs. Spalding, that the chief interest was directed—an interest that was founded in the Indian mind upon wonder, admiration, and awe, and in the minds of the trappers upon the powerful recollections awakened by seeing in their midst two refined Christian women, with the complexion and dress of their own mothers and sisters. United to this startling effect of memory, was respect for the religious devotion which had inspired them to undertake the long and dangerous journey to the Rocky Mountains, and also a sentiment of pity for what they knew only too well yet remained to be encountered by those delicate women in the prosecution of their duty.

Mrs. Whitman, who was in fine health, rode the greater part of the journey on horseback. She was a large, stately, fair-skinned woman, with blue eyes and light auburn, almost golden hair. Her manners were at once dignified and gracious. She was, both by nature and education a lady, and had a lady's appreciation of all that was courteous and refined, yet not without an element of romance and heroism in her disposition strong enough to have impelled her to undertake a missionary's life in the wilderness.

Mrs. Spalding was a different type of woman. Talented, and refined in her nature, she was less pleasing in exterior, and less attached to that which was superficially pleasing in others. But an indifference to outside appearances was in her case only a sign of her absorption in the work she had taken in hand. She possessed the true missionary spirit, and the talent to make it useful in an eminent degree, never thinking of herself, or the impression she made upon others, yet withal very firm and capable of command. Her health, which was always rather delicate, had suffered much from the fatigue of the journey, and the constant diet of fresh meat, and meat only, so that she was compelled at last to abandon horseback exercise, and to keep almost entirely to the light wagon of the missionaries.

As might be expected, the trappers turned from the contemplation of the pale, dark-haired occupant of the wagon, with all her humility and gentleness, to observe and admire the more striking figure, and more affably attractive manners of Mrs. Whitman. Meek, who never lost an opportunity to see and be seen, was seen riding alongside Mrs. Whitman, answering her curious inquiries, and entertaining her with stories of Blackfeet battles, and encounters with grizzly bears. Poor lady! Could she have looked into the future about which she was then so curious, she would have turned back appalled, and have fled with frantic fear to the home of her grieving parents. How could she then behold in the gay and boastful mountaineer, whose peculiarities of dress and speech so much diverted her, the very messenger who was to bear to the home of her girlhood the sickening tale of her bloody sacrifice to savage superstition and revenge? Yet so had fate decreed it.

When the trappers and Nez Percé had slaked their thirst for excitement by a few hours' travel in company with the Fur Company's and Missionary's caravan, they gave at length a parting display of horsemanship, and dashed off on the return trail to carry to camp the earliest news. It was on their arrival in camp that the Nez Percé and Flathead village, which had its encampment at the rendezvous ground on Green River[1], began to make preparations for the reception of the missionaries. It was then that Indian finery was in requisition! Then the Indian

---

1. This rendezvous of 1836, on the Green River near Horse Creek, is a celebrated one in the annals of the mountain trade because of this first appearance in the Rocky Mountains of white women. Mrs. Victor, in drawing the characters of these women and other missionaries, seems to have added the contributions of other memories to Meek's.

women combed and braided their long black hair, tying the plaits with gay-colored ribbons, and the Indian braves tied anew their streaming scalp locks, sticking them full of flaunting eagle's plumes, and not despising a bit of ribbon either. Paint was in demand both for the rider and his horse. Gay blankets, red and blue, buckskin fringed shirts, worked with beads and porcupine quills, and handsomely embroidered moccasins, were eagerly sought after. Guns were cleaned and burnished, and drums and fifes put in tune.

After a day of toilsome preparation, all was ready for the grand reception in the camp of the Nez Percé. Word was at length given that the caravan was in sight. There was a rush for horses, and in a few moments, the Indians were mounted and in line, ready to charge on the advancing caravan. When the command of the chiefs was given to start, a simultaneous chorus of yells and whoops burst forth, accompanied by the deafening din of the war drum, the discharge of firearms, and the clatter of the whole cavalcade, which was at once in a mad gallop toward the oncoming train. Nor did the yelling, whooping, drumming, and firing cease until within a few yards of the train.

All this demoniac hubbub was highly complimentary toward those for whom it was intended, but an unfortunate ignorance of Indian customs caused the missionaries to fail in appreciating the honor intended to them. Instead of trying to reciprocate the noise by an attempt at imitating it, the missionary camp was alarmed at the first burst and at once began to drive in their cattle and prepare for an attack. As the missionary party was in the rear of the train, they succeeded in getting together their loose stock before the Nez Percé had an opportunity of making themselves known, so that the leaders of the Fur Company, and Captain Stuart, had the pleasure of a hearty laugh at their expense, for the fright they had received.

A general shaking of hands followed the abatement of the first surprise, the Indian women saluting Mrs. Whitman and Mrs. Spalding with a kiss, and the missionaries were escorted to their camping ground near the Nez Percé encampment. Here, the whole village again formed in line, and a more formal introduction of the missionaries took place, after which they were permitted to go into camp.

When the intention of the Indians became known, Dr. Whitman, who was the leader of the missionary party, was boyishly delighted with the reception which had been given him. His frank, hearty, hopeful nature augured much good from the enthusiasm of the Indians. If his estimation of the native virtues of the savages was much too high, he suffered with those whom he caused to suffer for his belief, in the years which followed. Peace to the ashes of a good man! And honor to his associates, whose hearts were in the cause they had undertaken of Christianizing the Indians. Two of them still live—one of whom, Mr. Spalding, has conscientiously labored and deeply suffered for the faith. Mr. Gray, who was an unmarried man, returned the following year to the States, for a wife, and settled for a time among the Indians, but finally abandoned the missionary service, and removed to the Willamette Valley. These five persons constituted the entire force of teachers who could be induced at that time to devote their lives to the instruction of the savages in the neighborhood of the Rocky Mountains.

The trappers, and gentlemen of the fur company, and

Captain Stuart, had been passive but interested spectators of the scene between the Indians and the missionaries. When the excitement had somewhat subsided, and the various camps had become settled in their places, the tents of the white ladies were besieged with visitors, both civilized and savage. These ladies, who were making an endeavor to acquire a knowledge of the Nez Percé tongue in order to commence their instructions in the language of the natives, could have made very little progress, had their purpose been less strong than it was. Mrs. Spalding perhaps succeeded better than Mrs. Whitman in the difficult study of the Indian dialect. She seemed to attract the natives about her by the ease and kindness of her manner, especially the native women, who, seeing she was an invalid, clung to her rather than to her more lofty and self-asserting associate.

On the contrary, the leaders of the American Fur Company, Captain Wyeth and Captain Stuart, paid Mrs. Whitman the most marked and courteous attentions. She shone the bright particular star of that Rocky Mountain encampment, softening the hearts and the manners of all who came within her womanly influence. Not a gentleman among them but felt her silent command upon him to be his better self while she remained in his vicinity, and not a trapper or camp-keeper but respected the presence of womanhood and piety. But while the leaders paid court to her, the bashful trappers contented themselves with promenading before her tent. Should they succeed in catching her eye, they never failed to touch their beaver-skin caps in their most studiously graceful manner, though that should prove so dubious as to bring a mischievous smile to the blue eyes of the observant lady.

But our friend Joe Meek did not belong by nature to the bashful brigade. He was not content with disporting himself in his best trapper's toggery in front of a lady's tent. He became a not infrequent visitor, and amused Mrs. Whitman with the best of his mountain adventures, related in his soft, slow, yet smooth and firm utterance, and with many a merry twinkle of his mirthful dark eyes. In more serious moments he spoke to her of the future, and of his determination, sometime, to *settle down.* When she inquired if he had fixed upon any spot which in his imagination he could regard as *home,* he replied that he could not content himself to return to civilized life, but thought that when he gave up *bar fighting and Injun fighting* he should go down to the Willamette Valley and see what sort of life he could make of it there. How he lived up to this determination will be seen hereafter.

The missionaries remained at the rendezvous long enough to recruit their own strength and that of their stock, and to restore to something like health the invalid Mrs. Spalding, who, on changing her diet to dried meat, which the resident partners were able to supply her, commenced rapidly to improve. Letters were written and given to Capt. Wyeth to carry home to the States. The captain had completed his sale of Fort Hall and the goods it contained to the Hudson's Bay Company only a short time previous and was now about to abandon the effort to establish any enterprise either on the Columbia or in the Rocky Mountains. He had, however, executed his threat of the year previous, and punished the bad faith of the Rocky Mountain Company by placing them in direct competition with the Hudson's Bay Company.

The missionaries now prepared for their journey to the Columbia River. According to the advice of the mountain men, the heaviest wagon was left at the rendezvous, together with every heavy article that could be dispensed with. But Dr. Whitman refused to leave the light wagon, although assured he would never be able to get it to the Columbia, nor even to the Snake River. The good doctor had an immense fund of determination when there was an object to be gained or a principle involved. The only persons who did not oppose wagon transportation were the Indians. They sympathized with his determination, and gave him their assistance. The evidences of a different and higher civilization than they had ever seen were held in great reverence by them. The wagons, the domestic cattle, especially the cows and calves, were always objects of great interest with them. Therefore, they freely gave their assistance, and a sufficient number remained behind to help the Doctor, while the main party of both missionaries and Indians, having bidden the fur company and others farewell, proceeded to join the camp of two Hudson's Bay traders a few miles on their way.

The two traders, whose camp they now joined, were named McLeod and McKay. The latter, Thomas McKay, was the half-breed son of that unfortunate McKay in Mr. Astor's service, who perished on board the *Tonquin*, as related in Irving's Astoria. He was one of the bravest and most skillful partisans in the employ of the Hudson's Bay Company. McLeod had met the missionaries at the American rendezvous and invited them to travel in his company, an offer which they were glad to accept, as it secured them ample protection and other more trifling benefits, besides some society other than the Indians.

By dint of great perseverance, Dr. Whitman contrived to keep up with the camp day after day, though often coming in very late and very weary, until the party arrived at Fort Hall. At the fort the baggage was again reduced as much as possible, and Dr. Whitman was compelled by the desertion of his teamster to take off two wheels of his wagon and transform it into a cart which could be more easily propelled in difficult places. With this, he proceeded as far as the Boise River where the Hudson's Bay Company had a small fort or trading post, but here again, he was so strongly urged to relinquish the idea of taking his wagon to the Columbia, that after much discussion, he consented to leave it at Fort Boise until some future time when unencumbered by goods or passengers he might return for it.

Arrived at the crossing of the Snake River, Mrs. Whitman and Mrs. Spalding were treated to a new mode of ferriage, which even in their varied experience, they had never before met with. This new ferry was nothing more or less than a raft made of bundles of bulrushes woven together by grass ropes. Upon this frail flat-boat, the passengers were obliged to stretch themselves at length while an Indian swam across and drew it after him by a rope. As the waters of the Snake River are rapid and often *dancing mad*, it is easy to conjecture that the ladies were ill at ease on their bulrush ferry.

On went the party from the Snake River through the Grand Ronde to the Blue Mountains. The crossing here was somewhat difficult but accomplished in safety. The descent from the Blue Mountains on the west side gave the missionaries their first view of the country they had come to possess and to civilize and Christianize. That view was beautiful and grand—as goodly a prospect as longing eyes ever beheld this side of Canaan. Before them lay a country spread out like a map, with the windings of its rivers marked by fringes of trees, and its boundaries fixed by mountain ranges above which towered the snowy peaks of Mt. Hood, Mt. Adams, and Mt. Rainier. Far away could be traced the course of the Columbia, and over all the magnificent scene, glowed the red rays of sunset, tinging the distant blue of the mountains until they seemed shrouded in a veil of violet mist. It were not strange that with the reception given them by the Indians, and with this bird's-eye view of their adopted country, the hearts of the missionaries beat high with hope.

The descent from the Blue Mountains brought the party out on the Umatilla River, where they camped, Mr. McLeod parting company with them at this place to hasten forward to Fort Walla-Walla, and prepare for their reception. After two more days of slow and toilsome travel with cattle whose feet were cut and sore from the sharp rocks of the mountains, the company arrived safely at Walla-Walla fort, on the third of September. Here, they found Mr. McLeod, and Mr. Panbram who had charge of that post.

Mr. Panbram received the missionary party with every token of respect, and of pleasure at seeing ladies among them. The kindest attentions were lavished upon them from the first moment of their arrival, when the ladies were lifted from their horses, to the time of their departure. The apartments belonging to the fort being assigned to them, and all that the place afforded of comfortable living placed at their disposal. Here, for the first time in several months, they enjoyed the luxury of bread—a favor for which the suffering Mrs. Spalding was especially grateful.

At Walla-Walla the missionaries were informed that they

were expected to visit Vancouver, the headquarters of the Hudson's Bay Company on the Lower Columbia. After resting for two days, it was determined to make this visit before selecting places for mission work among the Indians. Accordingly, the party embarked in the company's boats, for the voyage down the Columbia, which occupied six days, owing to strong headwinds which were encountered at a point on the Lower Columbia, called Cape Horn. They arrived safely on the eleventh of September, at Vancouver, where they were again received with the warmest hospitality by the governor, Dr. John McLaughlin, and his associates. The change from the privations of wilderness life to the luxuries of Fort Vancouver was very great indeed, and two weeks passed rapidly away in the enjoyment of refined society, and all the other elegancies of the highest civilization.

At the end of two weeks, Dr. Whitman, Mr. Spalding, and Mr. Gray returned to the Upper Columbia, leaving the ladies at Fort Vancouver while they determined upon their several locations in the Indian country. After an absence of several weeks, they returned, having made their selections, and on the third day of November the ladies once more embarked to ascend the Columbia, to take up their residence in Indian wigwams while their husbands prepared rude dwellings by the assistance of the natives. The spot fixed upon by Dr. Whitman for his mission was on the Walla-Walla River about thirty miles from the fort of that name. It was called *Waiilatpu*, and the tribe chosen for his pupils were the Cayuses, a hardy, active, intelligent race, rich in horses and pasture lands.

Mr. Spalding selected a home on the Clearwater River, among the Nez Percé, of whom we already know so much. His mission was called *Lapwai*. Mr. Gray went among the Flatheads, an equally friendly tribe, and here we shall leave the missionaries to return to the Rocky Mountains and the life of the hunter and trapper. At a future date we shall fall in once more with these devoted people and learn what success attended their efforts to Christianize the Indians.

# 16

## 1836

The company of men who went north this year under Bridger and Fontenelle, numbered nearly three hundred. Rendezvous with all its varied excitements being over, this important brigade commenced its march. According to custom, the trappers commenced business on the headwaters of various rivers, following them down as the early frosts of the mountains forced them to do, until finally they wintered in the plains, at the most favored spots they could find in which to subsist themselves and animals.

From Green River, Meek proceeded with Bridger's command to Lewis River, Salt River, and other tributaries of the Snake, and camped with them in Pierre's Hole, that favorite mountain valley which every year was visited by the different fur companies.

Pierre's Hole, notwithstanding its beauties, had some repulsive features, or rather perhaps *one* repulsive feature, which was, its great numbers of rattlesnakes. Meek relates that being once caught in a very violent thunderstorm, he dismounted, and holding his horse, a fine one, by the bridle, himself took shelter under a narrow shelf of rock projecting from a precipitous bluff. Directly he observed an enormous rattlesnake hastening close by him to its den in the mountain. Congratulating himself on his snake-ship's haste to get out of the storm and his vicinity, he had only time to have one rejoicing thought when two or three others followed the trail of the first one. They were seeking the same rocky den, of whose proximity Meek now felt uncomfortably assured. Before these were out of sight, there came instead of twos and threes, tens and twenties, and then hundreds, and finally, Meek believes thousands, the ground being literally alive with them. Not daring to stir after he discovered the nature of his situation, he was obliged to remain and endure the disgusting and frightful scene, while he exerted himself to keep his horse quiet, lest the reptiles should attack him. By and by, when there were no more to come, but all were safe in their holes in the rock, Meek hastily mounted and galloped in the face of the tempest in preference to remaining longer in so unpleasant a neighborhood.

There was an old Frenchman among the trappers who used to charm rattlesnakes, and handling them freely, place them in his bosom, or allow them to wind about his arms, several at a time, their flat heads extending in all directions, and their bodies waving in the air, in the most snaky and nerve-shaking manner, to the infinite disgust of all the camp, and of Hawkins and Meek in particular. Hawkins often became so nervous that he threatened to shoot the Frenchman on the instant, if he did not desist, and great was the dislike he entertained for what he termed the *d—d infernal old wizard.*

It was often the case in the mountains and on the plains that the camp was troubled with rattlesnakes, so that each man on laying down to sleep found it necessary to encircle his bed with a hair rope, thus effectually fencing out the reptiles, which are too fastidious and sensitive of touch to crawl over a hair rope. But for this precaution, the trapper must often have shared his blanket couch with this foe to the *seed of the woman,* who, being asleep, would have neglected to *crush his head*, receiving instead the serpent's fang in *his heel*, if not in some nobler portion of his body.

There is a common belief abroad that the prairie dog harbors the rattlesnake, and the owl also, in his subterranean house, in a more or less friendly manner. Meek, however, who has had many opportunities of observing the habits of these three ill-assorted denizens of a common abode, gives it as his opinion that the prairie dog consents to the invasion of his premises alone through his inability to prevent it. As these prairie dog villages are always found on the naked prairies, where there is neither rocky den for the rattlesnake, nor shade for the blinking eyes of the owl, these two idle and impudent foreigners, availing themselves of the labors of the industrious little animal which builds itself a cool shelter from the sun, and a safe one from the storm, whenever their own necessities drive them to seek refuge from either sun or storm, enter uninvited and take possession. It is probable also, that so far from being a welcome guest, the rattlesnake occasionally gorges himself with a young prairie dog, when other game is not conveniently nigh, or that the owl lies in wait at the door of its borrowed-without-leave domicile, and succeeds in nabbing a careless field-mouse more easily than it could catch the same game by seeking it as an honest owl should do. The owl and the rattlesnake are like the Sioux when they go on a visit to the Omahas—the visit being always timed so as to be identical in date with that of the Government Agents who are distributing food and clothing. They are very good friends for the nonce, the poor Omahas not daring to be otherwise for fear of the ready vengeance on the next summer's buffalo hunt, therefore, they conceal their grimaces and let the Sioux eat them up, and when summer comes get massacred on their buffalo hunt, all the same.

But to return to our brigade. About the last of October Bridger's company moved down on to the Yellowstone by a circuitous route through the North Pass, now known as Hell Gate Pass, to Judith River, Mussel Shell River, Cross Creeks of the Yellowstone, Three Forks of Missouri, Missouri Lake, Beaver Head country, Big Horn River, and thence east again, and north again to the wintering ground in the great bend of the Yellowstone.

The company had not proceeded far in the Blackfeet country, between Hell Gate Pass and the Yellowstone, before they were attacked by the Blackfeet. On arriving at the Yellowstone they discovered a considerable encampment of the enemy on an island or bar in the river, and proceeded to open hostilities before the Indians should have discovered them. Making little forts of sticks or bushes, each man advanced cautiously to the bank overlooking the island, pushing his leafy fort before him as he crept silently nearer, until a position was reached whence firing could commence with effect. The first intimation the luckless savages had of the neighborhood of the whites was a volley of shots discharged into their camp, killing several of their number. But as this was their own mode of attack, no reflections were likely to be wasted upon the unfairness of the assault, quickly springing to their arms the firing was returned, and for several hours was kept up on both sides. At night, the Indians stole off, having lost nearly thirty killed. Nor did the trappers escape quite unhurt, three being killed and a few others wounded.[1]

Since men were of such value to the fur companies, it would seem strange that they should deliberately enter upon an Indian fight before being attacked. But unfortunate as these encounters really were, they knew of no other policy to be pursued. They—the American companies—were not resident, with a long acquaintance, and settled policy, such as rendered the Hudson's Bay Company so secure among the savages. They knew that among these unfriendly Indians, not to attack was to be attacked, and consequently, little time was ever given for an Indian to discover his vicinity to a trapper. The trapper's shot informed him of that, and afterward the race was to the swift, and the battle to the strong. Besides this acknowledged necessity for

1. Russell, in his account of the fall hunt of the Bridger brigade in 1836, says that Bridger divided his forces with instructions to meet on Yellowstone Lake. So it is possible that whatever little outfit Meek was with went around to Pierre's Hole and north from there into the present Park. But Joe's statement that they came to wintering ground at the bend of the Yellowstone—at present Livingston?—by such a northerly circuit flies in the face of probability. More likely the incident of Joe's adventure with Davis Crow and the surrounding events belong to this year, as Russell indicates—Russell, p. 41-51. Joe's recollections continue mostly to dovetail with others' if the reader mentally moves them forward one year. According to Russell, the Bridger men trapped the tributaries of the Yellowstone downstream from its great easterly turn and spent the winter on the Yellowstone near the mouth of Clark's Fork.

fighting whenever and wherever Indians were met with in the Blackfeet and Crow countries, almost every trapper had some private injury to avenge—some theft, or wound, or imprisonment, or at the very least, some terrible fright sustained at the hands of the universal foe. Therefore, there was no reluctance to shoot into an Indian camp, provided the position of the man shooting was a safe one, or more defensible than that of the man shot at. Add to this that there was no law in the mountains, only license, and it is easy to conjecture that might would have prevailed over right with far less incentive to the exercise of savage practices than actually did exist. Many a trapper undoubtedly shot his Indian *for the fun of it*, feeling that it was much better to do so than run the risk of being shot at for no better reason. Of this class of reasoners, it must be admitted, that Meek was one. Indian fighting, like bear fighting, had come to be a sort of pastime, in which he was proud to be known as highly accomplished. Having so many opportunities for the display of game qualities in encounters with these two by-no-means-to-be despised foes of the trapper, it was not often that they quarreled among themselves after the grand frolic of the rendezvous was over.

It happened, however, during this autumn, that while the main camp was in the valley of the Yellowstone, a party of eight trappers, including Meek and a comrade named Stanberry, were trapping together on the Mussel Shell, when the question as to which was the bravest man got started between them, and at length, in the heat of controversy, assumed such importance that it was agreed to settle the matter on the following day according to the Virginia code of honor, i.e., by fighting a duel, and shooting at each other with guns, which hitherto had only done execution on bears and Indians.

But some listening spirit of the woods determined to avert the danger from these two equally brave trappers, and save their ammunition for its legitimate use, by giving them occasion to prove their courage almost on the instant. While sitting around the campfire discussing the coming event of the duel at thirty paces, a huge bear, already wounded by a shot from the gun of their hunter who was out looking for game, came running furiously into camp, giving each man there a challenge to fight or fly.

"Now," spoke up one of the men quickly, "let Meek and Stanberry prove which is bravest, by fighting the bear!"

"Agreed," cried the two as quickly, and both sprang with guns and wiping-sticks in hand, charging upon the infuriated beast as it reached the spot where they were awaiting it. Stanberry was a small man, and Meek a large one. Perhaps it was owing to this difference of stature that Meek was first to reach the bear as it advanced. Running up with reckless bravado, Meek struck the creature two or three times over the head with his wiping stick before aiming to fire, which, however, he did so quickly and so surely that the beast fell dead at his feet. This act settled the vexed question. Nobody was disposed to dispute the point of courage with a man who would stop to strike a grizzly before shooting him, therefore, Meek was proclaimed by the common voice to be *cock of the walk* in that camp. The pipe of peace was solemnly smoked by himself and Stanberry, and the tomahawk buried never more to be resurrected between them, while a fat supper of bear meat celebrated the compact of everlasting amity.

It was not an infrequent occurrence for a grizzly bear to be run into camp by the hunters, in the Yellowstone country where this creature abounded. An amusing incident occurred not long after that just related, when the whole camp was at the Cross Creeks of the Yellowstone, on the south side of that river. The hunters were out, and had come upon two or three bears in a thicket. As these animals sometimes will do, they started off in a great fright, running toward camp, the hunters after them, yelling, frightening them still more. A runaway bear, like a runaway horse, appears not to see where it is going, but keeps right on its course no matter what dangers lie in advance. So one of these animals, having got headed for the middle of the encampment, saw nothing of what lay in its way, but ran on and on, apparently taking note of nothing but the yells in pursuit. So sudden and unexpected was the charge which he made upon camp, that the Indian women, who were sitting on the ground engaged in some ornamental work, had no time to escape out of the way. One of them was thrown down and run over, and another was struck with such violence that she was thrown twenty feet from the spot where she was hastily attempting to rise. Other objects in camp were upset and thrown out of the way, but without causing so much merriment as the mishaps of the two women who were so rudely treated by the monster.

It was also while the camp was at the Cross Creeks of the Yellowstone that Meek had one of his best-fought battles with a grizzly bear. He was out with two companions, one Gardiner, and Mark Head, a Shawnee Indian. Seeing a very large bear digging roots in the creek bottom, Meek proposed to attack it, if the others would hold his horse ready to mount if he failed to kill the creature. This being agreed to he advanced to within about forty paces of his game, when he raised his gun and attempted to fire, but the cap bursting, he only roused the beast, which turned on him with a terrific noise between a snarl and a growl, showing some fearful looking teeth. Meek turned to run for his horse, at the same time trying to put a cap on his gun, but when he had almost reached his comrades, their horses and his own took fright at the bear now close on his heels, and ran, leaving him alone with the now fully infuriated beast. Just at the moment, he succeeded in getting a cap on his gun, the teeth of the bear closed on his blanket capote which was belted around the waist, the suddenness and force of the seizure turning him around, as the skirt of his capote yielded to the strain and tore off at the belt. Being now nearly face to face with his foe, the intrepid trapper thrust his gun into the creature's mouth and attempted again to fire, but the gun being double triggered and not set, it failed to go off. Perceiving the difficulty, he managed to set the triggers with the gun still in the bear's mouth, yet no sooner was this done than the bear succeeded in knocking it out, and firing as it slipped out, it hit her too low down to inflict a fatal wound and only served to irritate her still farther.

In this desperate situation, when Meek's brain was rapidly working on the problem of live Meek or live bear, two fresh actors appeared on the scene in the persons of two cubs, who seeing their mother in difficulty, seemed desirous of doing something to assist her. Their appearance seemed to excite the bear to new exertions, for she made one desperate blow at Meek's empty gun with which he was defending himself, and knocked it out of his hands, and far down the bank or sloping hillside where the struggle was now going on. Then, being partially blinded by rage, she seized one of her cubs and began to box it about in a most unmotherly fashion. This diversion gave Meek a chance to draw his knife from the scabbard, with which he endeavored to stab the bear behind the ear, but she was too quick for him, and with a blow struck it out of his hand, as she had the gun, nearly severing his forefinger.

At this critical juncture, the second cub interfered, and got a boxing from the old bear, as the first one had done. This, too, gave Meek time to make a movement, and loosening his tomahawk from his belt, he made one tremendous effort, taking deadly aim, and struck her just behind the ear, the tomahawk sinking into the brain, and his powerful antagonist lay dead before him. When the blow was struck, he stood with his back against a little bluff of rock, beyond which it was impossible to retreat. It was his last chance, and his usual good fortune stood by him. When the struggle was over the weary victor mounted the rock behind him and looked down upon his enemy slain and *came to the conclusion that he was satisfied with bar fighting.*

But renown had sought him out even here, alone with his lifeless antagonist. Captain Stuart, with his artist, Mr. Miller[2], chanced upon this very spot, while yet the conqueror contemplated his slain enemy, and taking possession at once of the bear, whose skin was afterward preserved and stuffed, made a portrait of the *satisfied* slayer. A picture was subsequently painted by Miller of this scene, and was copied in wax for a museum in St. Louis, where it probably remains to this day, a monument of Meek's best bear fight. As for Meek's runaway horse and runaway comrades, they returned to the scene of action too late

2. Alfred Jacob Miller was not in the mountains at this time, but a year later he was, so Meek continues to date events a year early.

to be of the least service, except to furnish our hero with transportation to camp, which, considering the weight of his newly gathered laurels, was no light service after all.

In November, Bridger's camp arrived at the Bighorn River, expecting to winter, but finding the buffalo all gone, were obliged to cross the mountains lying between the Bighorn and Powder rivers to reach the buffalo country on the latter stream. The snow having already fallen quite deep on these mountains, the crossing was attended with great difficulty, and many horses and mules were lost by sinking in the snow, or falling down precipices made slippery by the melting and freezing of the snow on the narrow ridges and rocky benches along which they were forced to travel.

About Christmas, all the company went into winter quarters on Powder River, in the neighborhood of a company of Bonneville's men, left under the command of Antoine Montero, who had established a trading post[3] and fort at this place, hoping, no doubt, that here they should be comparatively safe from the injurious competition of the older companies. The appearance of three hundred men, who had the winter before them in which to do mischief, was therefore as unpleasant as it was unexpected, and the result proved that even Montero, who was Bonneville's experienced trader, could not hold his own against so numerous and expert a band of marauders as Bridger's men, assisted by the Crows, proved themselves to be, for by the return of spring Montero had very little remaining of the property belonging to the fort, nor anything to show for it.

3. This post, called Portuguese Houses, was built near the present Kaycee, Wyoming in 1834 by the Portuguese partisan for Bonneville, Antonio Montero.

This mischievous war upon Bonneville was prompted partly by the usual desire to cripple a rival trader, which the leaders encouraged in their men, but in some individual instances, far more by the desire for revenge upon Bonneville personally, on account of his censures passed upon the members of the Monterey expedition, and on the ways of mountain men generally.

About the first of January, Fontenelle, with four men, and Captain Stuart's party, left camp to go to St. Louis for supplies. At Fort Laramie, Fontenelle committed suicide[4], in a fit of *mania a potu*, and his men returned to camp with the news.

4. Fontenelle lived beyond this reported suicide—Father DeSmet reports giving this partisan the last rites a year later—Hafen series, V, p. 99. Yet William Drummond Stewart's autobiographical narrator in Edward Warren also reports Fontenelle's suicide. DeVoto—p. 439—says that Joe was repeating a story commonly told in the mountains. An anomaly.

# 17

## 1837

The fate of Fontenelle should have served as a warning to his associates and fellows. *Should have done,* however, are often idle words, and as sad as they are idle, they match the poets *might have been,* in their regretful impotency. Perhaps there never was a winter camp in the mountains more thoroughly demoralized than that of Bridger during the months of January and February. Added to the whites, who were reckless enough, were a considerable party of Delaware and Shawnee Indians, excellent allies, and skillful hunters and trappers, but having the Indian's love of strong drink. "Times were pretty good in the mountains," according to the mountain man's notion of good times. That is to say, beaver was plenty, camp large, and alcohol abundant, if dear. Under these favorable circumstances, much alcohol was consumed, and its influence was felt in the manners not only of the trappers, white and red, but also upon the neighboring Indians.

The Crows, who had for two years been on terms of a sort of semi-amity with the whites, found it to their interest to conciliate so powerful an enemy as the American Fur Company was now become, and made frequent visits to the camp, on which occasion they usually succeeded in obtaining a taste of the fire-water of which they were inordinately fond. Occasionally a trader was permitted to sell liquor to the whole village, when a scene took place whose peculiar horrors were wholly indescribable, from the inability of language to convey an adequate idea of its hellish degradation. When a trader sold alcohol to a village it was understood both by himself and the Indians what was to follow. And to secure the trader against injury, a certain number of warriors were selected out of the village to act as a police force and to guard the trader during the *drunk* from the insane passions of his customers. To the police not a drop was to be given.

This being arranged, and the village disarmed, the carousal began. Every individual, man, woman, and child, was permitted to become intoxicated. Every form of drunkenness, from the simple stupid to the silly, the heroic, the insane, the beastly, the murderous, displayed itself. The scenes which were then enacted beggared description, as they shocked the senses of even the hard-drinking, license-loving trappers who witnessed them. That they did not *point a moral* for these men, is the strangest part of the whole transaction.

When everybody, police excepted, was drunk as drunk could be, the trader began to dilute his alcohol with water, until finally, his keg contained water only, slightly flavored by the washings of the keg, and as they continued to drink of it without detecting its weak quality, they finally drank themselves sober, and were able at last to sum up the cost of their intoxication. This was generally nothing less than the whole property of the village, added to which were not a few personal injuries and usually a few murders. The village now being poor, the Indians were correspondingly humble and were forced to begin a system of reprisal by stealing and making war, a course for which the traders were prepared and which they avoided by leaving that neighborhood. Such were some of the sins and sorrows for which the American fur companies were answerable and which detracted seriously from the respect that the courage and other good qualities of the mountain men freely commanded.

By the first of March, these scenes of wrong and riot were over, for that season at least, and camp commenced moving back toward the Blackfoot country. After recrossing the mountains, passing the Bighorn, Clarke's, and Rosebud Rivers, they came upon a Blackfoot village on the Yellowstone, which as usual they attacked, and a battle ensued, in which Manhead, captain of the Delawares was killed, another Delaware named Tom Hill succeeding him in command. The fight did not result in any great loss or gain to either party. The camp of Bridger fought its way past the village, which was what they must do, in order to proceed.

Meek, however, was not quite satisfied with the punishment the Blackfeet had received for the killing of Manhead, who had been in the fight with him when the Camanches attacked them on the plains. Desirous of doing something on his own account, he induced a comrade named LeBlas, to accompany him to the village, after night had closed over the scene of the late contest.

Stealing into the village with a noiselessness equal to that of one of Fennimore Cooper's Indian scouts, these two daring trappers crept so near that they could look into the lodges, and see the Indians at their favorite game of *Hand*. Inferring from this that the savages did not feel their losses very severely, they determined to leave some sign of their visit, and wound their enemy in his most sensitive part, the horse. Accordingly, they cut the halters of a number of the animals, fastened in the customary manner to a stake, and succeeded in getting off with nine of them, which property they proceeded to appropriate to their own use.

As the spring and summer advanced, Bridger's brigade advanced into the mountains, passing the Cross Creek of the Yellowstone, Twenty-Five-Yard River, Cherry River, and coming on to the headwaters of the Missouri spent the early part of the summer in that locality. Between Gallatin and Madison forks, the camp struck the great trail of the Blackfeet. Meek and Mark Head had fallen four or five days behind camp, and being on this trail felt a good deal of uneasiness. This feeling was not lessened by seeing, on coming to Madison Fork, the skeletons of two men tied to or suspended from trees, the flesh eaten off their bones. Concluding discretion to be the safest part of valor in this country, they concealed themselves by day and traveled by night, until camp was finally reached near Henry's Lake. On this march, they forded a flooded river, on the back of the same mule, their traps placed on the other, and escaped from pursuit of a dozen yelling savages, who gazed after them in astonishment. "Taking their mule," said Mark Head, "to be a beaver, and themselves great medicine men."

"That," said Meek, "is what I call *cooning* a river."

From this point, Meek set out with a party of thirty or forty trappers to travel up the river to headwaters, accompanied by the famous Indian painter Stanley, whose party was met with, this spring, traveling among the mountains. The party of trappers were a day or two ahead of the main camp when they found themselves following close after the big Blackfoot village which had recently passed over the trail, as could be seen by the usual signs and also by the dead bodies strewn along the trail, victims of that horrible scourge, the smallpox. The village was evidently fleeing to the mountains, hoping to rid itself of the plague in their colder and more salubrious air.

Not long after coming upon these evidences of proximity to an enemy, a party of a hundred and fifty of their warriors were discovered encamped in a defile or narrow bottom enclosed by high bluffs, through which the trappers would have to pass. Seeing that in order to pass this war party and the village, which was about half a mile in advance, there would have to be some fighting done, the trappers resolved to begin the battle at once by attacking their enemy, who was as yet ignorant of their neighborhood. In pursuance of this determination, Meek, Newell, Mansfield, and Le Blas, commenced hostilities. Leaving their horses in camp, they crawled along on the edge of the overhanging bluff until opposite to the encampment of Blackfeet, firing on them from the shelter of some bushes which grew among the rocks. But the Blackfeet, though ignorant of the number of their enemy, were not to be dislodged so easily, and after an hour or two of random shooting, contrived to scale the bluff at a point higher up, and to get upon a ridge of ground still higher than that occupied by the four trappers. This movement dislodged the latter, and they hastily retreated through the bushes and returned to camp.

The next day, the main camp having come up, the fight was renewed. While the greater body of the company, with the pack-horses, were passing along the high bluff overhanging them, the party of the day before, and forty or fifty others, undertook to drive the Indians out of the bottom, and by keeping them engaged, allow the train to pass in safety. The trappers rode to the fight on this occasion, and charged the Blackfeet furiously, they having joined the village a little farther on. A general skirmish now took place. Meek, who was mounted on a fine horse, was in the thickest of the fight. He had at one time a side-to-side race with an Indian who strung his bow so hard that the arrow dropped, just as Meek, who had loaded his gun running, was ready to fire, and the Indian dropped after his arrow.

Newell, too, had a desperate conflict with a half-dead warrior, who having fallen from a wound, he thought dead and was trying to scalp. Springing from his horse, he seized the Indian's long, thick hair in one hand and with his knife held in the other, made a pass at the scalp, when the savage roused up knife in hand, and a struggle took place in which it was for a time doubtful which of the combatants would part with the coveted scalp-lock. Newell might have been glad to resign the trophy, and leave the fallen warrior his tuft of hair, but his fingers were in some way caught by some gun-screws with which the savage had ornamented his *coiffure,* and would not part company. In this dilemma, there was no other alternative but to fight. The miserable savage was dragged a rod or two in the struggle, and finally dispatched.

Mansfield also got into such close quarters, surrounded by

the enemy, that he gave himself up for lost, and called out to his comrades: “Tell old Gabe—Bridger—that old Cotton—his own sobriquet—is gone.” He lived, however, to deliver his own farewell message, for at this critical juncture the trappers were reenforced and relieved. Still, the fight went on, the trappers gradually working their way to the upper end of the enclosed part of the valley, past the point of danger.

Just before getting clear of this entanglement Meek became the subject of another picture, by Stanley, who was viewing the battle from the heights above the valley.[1] The picture which is well known as *The Trapper's Last Shot*, represents him as he turned upon his horse, a fine and spirited animal, to discharge his last shot at an Indian pursuing, while in the bottom, at a little distance away, other Indians are seen skulking in the tall reedy grass.

The last shot having been discharged with fatal effect, our trapper, so persistently lionized by painters, put his horse to his utmost speed and soon after overtook the camp, which had now passed the strait of danger. But, the Blackfeet were still unsatisfied with the result of the contest. They followed after, reinforced from the village, and attacked the camp. In the fight which followed, a Blackfoot woman's horse was shot down, and Meek tried to take her prisoner, but two or three of her people coming to the rescue, engaged his attention, and the woman was saved by seizing hold of the tail of her husband's horse, which setting off at a run, carried her out of danger.

---

1. DeVoto seems to have established that Meek did not meet Stanley until 1847, and that Stanley did no painting called The Trapper's Last Shot, though he did paint Joe Meek.

The Blackfeet found the camp of Bridger too strong for them.[2] They were severely beaten and compelled to retire to their village, leaving Bridger free to move on. The following day, the camp reached the village of Little-Robe, a chief of the Peagans, who held a talk with Bridger, complaining that his nation were all perishing from the smallpox which had been given to them by the whites. Bridger was able to explain to Little-Robe his error, inasmuch as although the disease might have originated among the Whites, it was communicated to the Blackfeet by Jim Beckwith, a negro, and principal chief of their enemies the Crows. This unscrupulous wretch had caused two infected articles to be taken from a Mackinaw boat, up from St. Louis, and disposed of to the Blackfeet—whence the horrible scourge under which they were suffering.

This matter being explained, Little-Robe consented to trade horses and skins, and the two camps parted amicably. The next day after this friendly talk, Bridger being encamped on the trail in advance of the Blackfeet, an Indian came riding into camp with his wife and daughter, packhorse and lodge pole, and all his worldly goods, unaware until he got there of the snare into which he had fallen. The French trappers, generally, decreed to kill the man and take possession of the woman. But Meek, Kit Carson, and others of the American trappers of the better sort, interfered to prevent this truly savage act. Meek took the woman's horse by the head, Carson the man's, the daughter following, and led them out of camp. Few of the Frenchmen cared to interrupt either of these two men, and they were suffered to depart in peace. When at a safe distance, Meek

---

2. Russell also tells the tale of this fight, dating it June, 1838.

stopped, and demanded as some return for having saved the man's life, a present of tobacco, a luxury which, from the Indian's pipe, he suspected him to possess. About enough for two chews was the result of this demand, complied with rather grudgingly, the Indian viewing with the trapper in his devotion to the weed. Just at this time, owing to the death of Fontenelle, and a consequent delay in receiving supplies, tobacco was scarce among the mountaineers.

Bridger's brigade of trappers met with no other serious interruptions on their summer's march. They proceeded to Henry's Lake, and crossing the Rocky Mountains, traveled through the Pine Woods, always a favorite region, to Lewis's Lake on Lewis's Fork of the Snake River, and finally up the Grovant Fork, recrossing the mountains to Wind River, where the rendezvous for this year was appointed.[3]

Here, once more, the camp was visited by a last years' acquaintance. This was none other than Mr. Gray[4], of the Flathead Mission, who was returning to the States on business connected with the missionary enterprise, and to provide himself with a helpmeet for life—a co-laborer and sufferer in the contemplated toil of teaching savages the rudiments of a religion difficult even to the comprehension of an old civilization.

Mr. Gray was accompanied by two young men—whites—

---

3. Joe has his rendezvous confused. The Gray party was returning to the States in 1837, when rendezvous was on Green River. It returned westward the next year, when the get-together was at the mouth of the Popo Agie—present Riverton, Wyoming.

4. Mrs. Victor apparently got some of her information on the Gray affair from Gray or Gray's journal or his History of Oregon-Portland, 1870. The journal—(Whitman College Quarterly, June, 1943—identifies the French trader who intervened as Papair.

who wished to return to the States, and also by a son of one of the Flathead chiefs. Two other Flathead Indians, and one Iroquois and one Snake Indian, were induced to accompany Mr. Gray. The undertaking was not without danger, and so the leaders of the fur company assured him. But Mr. Gray was inclined to make light of the danger, having traveled with entire safety when under the protection of the Fur Companies the year before. He proceeded without interruption until he reached Ash Hollow, in the neighborhood of Fort Laramie, when his party was attacked by a large band of Sioux, and compelled to accept battle. The five Indians, with the whites, fought bravely, killing fifteen of the Sioux, before a parley was obtained by the intervention of a French trader who chanced to be among the Sioux. When Mr. Gray was able to hold a *talk* with the attacking party, he was assured that his life and that of his two white associates would be spared, but that they wanted to kill the strange Indians and take their fine horses. It is not at all probable that Mr. Gray consented to this sacrifice, though he has been accused of doing so.

No doubt the Sioux took advantage of some hesitation on his part, and rushed upon his Indian allies in an unguarded moment. However that may be, his allies were killed and he was allowed to escape, after giving up the property belonging to them, and a portion of his own.

This affair was the occasion of much ill-feeling toward Mr. Gray, when, in the following year, he returned to the mountains with the tale of the massacre of his friends and his own escape. The mountain men, although they used their influence to restrain the vengeful feelings of the Flathead tribe, whispered among themselves that Gray had preferred his own life to that of his friends. The old Flathead chief, too, who had lost a son by the massacre, was hardly able to check his impulsive desire for revenge, for he held Mr. Gray responsible for his son's life. Nothing more serious, however, grew out of this unhappy tragedy than a disaffection among the tribe toward Mr. Gray, which made his labors useless, and finally determined him to move to the Willamette Valley.

There were no outsiders besides Gray's party at the rendezvous of this year, except Captain Stuart, and he was almost as good a mountaineer as any. This doughty English traveler had the bad fortune, together with that experienced leader Fitzpatrick, of being robbed by the Crows in the course of the fall hunt, in the Crow country. These expert horse thieves had succeeded in stealing nearly all the horses belonging to the joint camp, and had so disabled the company that it could not proceed. In this emergency, Newell, who had long been a subtrader and was wise in Indian arts and wiles, was sent to hold a talk with the thieves. The talk was held, according to custom, in the Medicine lodge, and the usual amount of smoking, of long silences, and grave looks, had to be participated in, before the subject on hand could be considered. Then the chiefs complained as usual of wrongs at the hands of the white men, of their fear of smallpox, from which some of their tribe had suffered, of friends killed in battle with the whites, and all the list of ills that Crow flesh is heir to at the will of their white enemies. The women, too, had their complaints to proffer, and the number of widows and orphans in the tribe was pathetically set forth. The chiefs also made a strong point of this latter complaint, and on it the wily Newell hung his hopes of recovering the stolen property.

"It is true," said he to the chiefs, "that you have sustained heavy losses. But that is not the fault of the Blanket chief—Bridger. If your young men have been killed, they were killed when attempting to rob or kill our captain's men. If you have lost horses, your young men have stolen five to our one. If you are poor in skins and other property, it is because you sold it all for drink which did you no good. Neither is Bridger to blame that you have had the smallpox. Your own chief, in trying to kill your enemies, the Blackfeet, brought that disease into the country.

"But it is true that you have many widows and orphans to support, and that is bad. I pity the orphans, and will help you to support them, if you will restore to my captain the property stolen from his camp. Otherwise, Bridger will bring more horses, and plenty of ammunition, and there will be more widows and orphans among the Crows than ever before."[5]

This was a kind of logic that was easy to understand and quick to convince among savages. The bribe, backed by a threat, settled the question of the restoration of the horses, which were returned without further delay, and a present of blankets and trinkets was given, ostensibly to the bereaved women, really to the covetous chiefs.

5. This incident, as reported by Meek, differs in too many details from other similar incidents for the reports of them to be considered corroboration.

# 18

**1837**

The decline of the business of hunting furs began to be quite obvious about this time. Besides the American and St. Louis Companies, and the Hudson's Bay Company, there were numerous lone traders with whom the ground was divided. The autumn of this year was spent by the American Company, as formerly, in trapping beaver on the streams issuing from the eastern side of the Rocky Mountains. When the cold weather finally drove the fur company to the plains, they went into winter quarters once more in the neighborhood of the Crows on Powder River. Here were re-enacted the wild scenes of the previous winter, both trappers and Indians being given up to excesses.

On the return of spring, Bridger again led his brigade all through the Yellowstone country, to the streams on the north side of the Missouri, to the headwaters of that river, and finally rendezvoused on the north fork of the Yellowstone, near Yellowstone Lake.[1] Though the amount of furs taken on the spring hunt was considerable, it was by no means equal to former years. The fact was becoming apparent that the beaver was being rapidly exterminated.

However there was beaver enough in camp to furnish the means for the usual profligacy. Horse-racing, betting, gambling, and drinking, were freely indulged in. In the midst of this *fun*, there appeared at the rendezvous Mr. Gray, now accompanied by Mrs. Gray and six other missionary ladies and gentlemen. Here also were two gentlemen from the Methodist mission on the Wallamet, who were returning to the States. Captain Stuart was still traveling with the fur company, and was also present with his party, besides which a Hudson's Bay trader named Ematinger was encamped nearby. As if actuated to extraordinary displays by the unusual number of visitors, especially the four ladies, both trappers and Indians conducted themselves like the madcaps they were. The Shawnees and Delawares danced their great war-dance before the tents of the missionaries, and Joe Meek, not to be outdone, arrayed himself in a suit of armor belonging to Captain Stuart and strutted about the encampment, then mounting his horse, played the part of an ancient knight, with a good deal of *eclat*.

Meek had not abstained from the alcohol kettle, but had offered it and partaken of it rather more freely than usual, so

---

1. Joe's statement about holding rendezvous near Yellowstone Lake is confusing. Perhaps various small outfits making up Bridger's large brigade did meet near Yellowstone earlier—Russell's account would be consistent with that. But the formal summer rendezvous, with the appearance of the Gray party, took place at the mouth of the Popo Agie.

that when rendezvous was broken up, the St. Louis Company gone to the Popo Agie, and the American Company going to Wind River, he found that his wife, a Nez Percé who had succeeded Umentucken in his affections, had taken offense, or a fit of homesickness, which was synonymous, and departed with the party of Ematinger and the missionaries, intending to visit her people at Walla-Walla. This desertion wounded Meek's feelings, for he prided himself on his courtesy to the sex, and did not like to think that he had not behaved handsomely. All the more was he vexed with himself because his spouse had carried with her a pretty and sprightly baby daughter, of whom the father was fond and proud, and who had been christened Helen Mar[2], after one of the heroines of Miss Porter's *Scottish Chiefs*—a book much admired in the mountains, as it has been elsewhere.

Therefore at the first camp of the American Company, Meek resolved to turn his back on the company, and go after the mother and daughter. Obtaining a fresh kettle of alcohol, to keep up his spirits, he left camp, returning toward the scene of the late rendezvous. But in the effort to keep up his spirits he had drank too much alcohol, and the result was that on the next morning, he found himself alone on the Wind River Mountain, with his horses and pack mules, and very sick indeed. Taking a little more alcohol to brace up his nerves, he started on again, passing around the mountain on to the Sweetwater, thence to the Sandy, and thence across a country without water for seventy-five miles, to Green River, where the camp of Ematinger was overtaken.

---

2. Helen Mar Meek was killed in the Whitman massacre by the Cayuse Indians at Waiilatpu Mission in 1847.

The heat was excessive and the absence of water made the journey across the arid plain between Sandy and Green Rivers one of great suffering to the traveler and his animals, and the more so as the frequent references to the alcohol kettle only increased the thirst-fever instead of allaying it. But Meek was not alone in suffering. About halfway across the scorching plain, he discovered a solitary woman's figure standing in the trail, and two riding horses near her, whose drooping heads expressed their dejection. On coming up with this strange group, Meek found the woman to be one of the missionary ladies, a Mrs. Smith, and that her husband was lying on the ground, dying, as the poor sufferer believed himself, for water.

Mrs. Smith made a weeping appeal to Meek for water for her dying husband, and truly the poor woman's situation was a pitiable one. Behind camp, with no protection from the perils of the desert and wilderness—only a terrible care instead—the necessity of trying to save her husband's life. As no water was to be had, alcohol was offered to the famishing man, who, however, could not be aroused from his stupor of wretchedness. Seeing that death really awaited the unlucky missionary unless something could be done to cause him to exert himself, Meek commenced at once, and with unction, to abuse the man for his unmanliness. His style, though not very refined, was certainly very vigorous.

"You're a d—d pretty fellow to be lying on the ground here, lolling your tongue out of your mouth, and trying to die. Die, if you want to, and to hell with you, you'll never be missed. Here's your wife, who you are keeping standing here in the hot sun. Why don't *she* die? She's got more pluck than a white-livered chap like you. But I'm not going to leave her waiting here for you to die. Thar's a band of Indians behind on the trail, and I've been riding like hell to keep out of their way. If you want to stay here and be scalped, you can stay. Mrs. Smith is going with me. Come, madam," continued Meek, leading up her horse. "Let me help you to mount, for we must get out of this cursed country as fast as possible."

Poor Mrs. Smith did not wish to leave her husband, nor did she relish the notion of staying to be scalped. Despair tugged at her heartstrings. She would have sunk to the ground in a passion of tears, but Meek was too much in earnest to permit precious time to be thus wasted. "Get on your horse," said he rather roughly. "You can't save your husband by staying here, crying. It is better that one should die than two, and he seems to be a worthless dog anyway. Let the Indians have him."

Almost lifting her upon the horse, Meek tore the distracted woman away from her husband, who had yet strength enough to gasp out an entreaty not to be left.

"You can follow us if you choose," said the apparently merciless trapper, "or you can stay where you are. Mrs. Smith can find plenty of better men than you. Come, madam!" and he gave the horse a stroke with his riding-whip which started him into a rapid pace.

The unhappy wife, whose conscience reproached her for leaving her husband to die alone, looked back, and saw him raising his head to gaze after them. Her grief broke out afresh, and she would have gone back even then to remain with him, but Meek was firm, and again started up her horse. Before they were quite out of sight, Meek turned in his saddle, and beheld the dying man sitting up. "Hurrah," said he, "he's all right. He will overtake us in a little while." And as he predicted, in little over an hour, Smith came riding up, not more than half dead by this time. The party got into camp on Green River, about eleven o'clock that night, and Mrs. Smith, having told the story of her adventures with the unknown trapper who had so nearly kidnapped her, the laugh and the cheer went round among the company. "That's Meek," said Ematinger, "you may rely on that. He's just the one to kidnap a woman in that way." When Mrs. Smith fully realized the service rendered, she was abundantly grateful, and profuse were the thanks which our trapper received, even from the much-abused husband, who was now thoroughly alive again. Meek failed to persuade his wife to return with him. She was homesick for her people, and would go to them. But instead of turning back, he kept on with Ematinger's camp as far as Fort Hall, which post was then in charge of Courtenay Walker.

While the camp was at Soda Springs, Meek observed the missionary ladies baking bread in a tin reflector before a fire. Bread was a luxury unknown to the mountain man, and as a sudden recollection of his boyhood, and the days of bread-and-butter came over him, his mouth began to water. Almost against his will he continued to hang round the missionary camp, thinking about the bread. At length, one of the Nez Perces, named James, whom the missionary had taught to sing, at their request struck up a hymn, which he sang in a very creditable manner. As a reward of his pious proficiency, one of the ladies gave James a biscuit. A bright thought struck our longing hero's brain. "Go back," said he to James, "and sing another hymn, and when the ladies give you another biscuit, bring it to me." And in this manner, he obtained a taste of the coveted luxury, bread—of which, during nine years in the mountains he had not eaten.

At Fort Hall, Meek parted company with the missionaries, and with his wife and child. As the little black-eyed daughter took her departure in company with this new element in savage life—the missionary society—her father could have had no premonition of the fate to which the admixture of the savage and the religious elements was step by step consigning her.

After remaining a few days at the fort, Meek, who found some of his old comrades at this place, went trapping with them up the Portneuf, and soon made up a pack of one hundred and fifty beaver skins. These, on returning to the fort, he delivered to Jo. Walker, one of the American Company's traders at that time, and took Walker's receipt for them. He then, with Mansfield and Wilkins, set out about the first of September for the Flathead country, where Wilkins had a wife. In their company was an old Flathead woman, who wished to return to her people, and took this opportunity.

The weather was still extremely warm. It had been a season of great drought, and the streams were nearly all entirely dried up. The first night out, the horses, eight in number, strayed off in search of water, and were lost. Now commenced a day of fearful sufferings. No water had been found since leaving the fort. The loss of the horses made it necessary for the company to separate to look for them. Mansfield and Wilkins going in one direction, Meek and the old Flathead woman in another. The little coolness and moisture which night had imparted to the atmosphere was quickly dissipated by the unchecked rays of the pitiless sun shining on a dry and barren plain, with not a vestige of verdure anywhere in sight. On and on went the old Flathead woman, keeping always in the advance, and on and on followed Meek, anxiously scanning the horizon for a chance sight of the horses.

Higher and higher mounted the sun, the temperature increasing in intensity until the great plain palpitated with radiated heat, and the horizon flickered almost like a flame where the burning heavens met the burning earth. Meek had been drinking a good deal of rum at the fort, which circumstance did not lessen the terrible consuming thirst that was torturing him.

Noon came, and passed, and still the heat and the suffering increased, the fever and craving of hunger being now added to that of thirst. On and on, through the whole of that long scorching afternoon, trotted the old Flathead woman in the peculiar traveling gait of the Indian and the mountaineer, Meek following at a little distance, and going mad, as he thought, for a little water. And mad he probably was, as famine sometimes makes its victims. When night at last closed in, he laid down to die, as the missionary Smith had done before. But he did not remember Smith: he only thought of water, and heard it running, and fancied the old woman was lapping it like a wolf. Then he rose to follow her and find it, it was always just ahead, and the woman was howling to him to show him the trail.

Thus the night passed, and in the cool of the early morning, he experienced a little relief. He was really following his guide, who as on the day before, was trotting on ahead. Then the thought possessed him to overtake and kill her, hoping from her shriveled body to obtain a morsel of food, and drop of moisture. But his strength was failing, and his guide so far ahead that he gave up the thought as involving too great exertion, continuing to follow her in a helpless and hopeless kind of way.

At last! There was no mistake this time, he heard running water, and the old woman *was* lapping it like a wolf. With a shriek of joy, he ran and fell on his face in the water, which was not more than one foot in depth, nor the stream more than fifteen feet wide. But it had a white pebbly bottom, and the water was clear, if not very cool. It was something to thank God for, which the none too religious trapper acknowledged by a fervent "Thank God!"

For a long time, he lay in the water, swallowing it, and by thrusting his finger down his throat, vomiting it up again, to prevent surfeit, his whole body taking in the welcome moisture at all its million pores. The fever abated, a feeling of health returned, and the late perishing man was restored to life and comparative happiness. The stream proved to be Godin's Fork, and here Meek and his faithful old guide rested until evening, in the shade of some willows, where their good fortune was completed by the appearance of Mansfield and Wilkins with the horses. The following morning, the men found and killed a fat buffalo cow, whereby all their wants were supplied, and good feeling restored in the little camp.

From Godin's Fork they crossed over to Salmon River, and presently struck the Nez Percé trail which leads from that river over into the Beaver-head country, on the Beaver-head or Jefferson Fork of the Missouri, where there was a Flathead and Nez Percé village, on or about the present site of Virginia City, in Montana.

Not stopping long here, Meek and his companions went on to the Madison Fork with the Indian village, and to the shores of Missouri Lake, joining in the fall hunt for buffalo.

# 19

"Tell me all about a buffalo hunt," said the writer to Joe Meek, as we sat at a window overlooking the Columbia River, where it has a beautiful stretch of broad waters and curving wooded shores, and talking about mountain life. "Tell me how you used to hunt buffalo."

"Waal, there is a good deal of sport in runnin' buffalo. When the camp discovered a band, then every man that wanted to run, made haste to catch his buffalo horse. We sometimes went out thirty or forty strong, sometimes two or three, and at other times a large party started on the hunt, the more the merrier. We alway had great bantering about our horses, each man, according to his own account, having the best one.

"When we first start, we ride slow, so as not to alarm the buffalo. The nearer we come to the band the greater our excitement. The horses seem to feel it too, and are worrying to be off. When we come so near that the band starts, then the word is given, our horses' mettle is up, and away we go!

"Thar may be ten thousand in a band. Directly, we crowd them so close that nothing can be seen but dust, nor anything heard but the roar of their trampling and bellowing. The hunter now keeps close on their heels to escape being blinded by the dust, which does not rise as high as a man on horseback, for thirty yards behind the animals. As soon as we are close enough, the firing begins, and the band is on the run, and a herd of buffalo can run about as fast as a good race-horse. How they *do* thunder along! They give us a pretty sharp race. Take care! Down goes a rider, and away goes his horse with the band. Do you think we stopped to look after the fallen man? Not we. We rather thought that war fun, and if he got killed, why, 'he war unlucky, that war all. Plenty more men: couldn't bother about him.'

"Thar's a fat cow ahead. I force my way through the band to come up with her. The buffalo crowd around so that I have to put my foot on them, now on one side, now the other, to keep them off my horse. It is lively work, I can tell you. A man has to look sharp not to be run down by the band pressing him on, buffalo and horse at the top of their speed.

"Look out, thar's a ravine ahead, as you can see by the plunge which the band makes. Hold up! If the band is large, it fills the ravine full to the brim, and the hindmost of the herd pass over on top of the foremost. It requires horsemanship not to be carried over without our own consent, but then we mountain men are *all* good horsemen. Over the ravine we go, but we do it our own way.

"We keep up the chase for about four miles, selecting our game as we run, and killing a number of fat cows to each man, some more and some less. When our horses are tired, we slacken up, and turn back. We meet the camp-keepers with packhorses.

They soon butcher, pack up the meat, and we all return to camp, whar we laugh at each other's mishaps, and eat fat meat: and this constitutes the glory of mountain life."

"But you were going to tell me about the buffalo hunt at Missouri Lake?"

"Thar isn't much to tell. It war pretty much like other buffalo hunts. Thar war a lot of us trappers happened to be at a Nez Percé and Flathead village in the fall of '38, when they war agoin' to kill winter meat, and as their hunt lay in the direction we war going, we joined in. The old Nez Percé chief, *Kow-e-so-te* had command of the village, and we trappers had to obey him, too.

"We started off slow, nobody war allowed to go ahead of camp. In this manner, we caused the buffalo to move on before us, but not to be alarmed. We war eight or ten days traveling from the Beaver-head to Missouri Lake, and by the time we got than, the whole plain around the lake war crowded with buffalo, and it war a splendid sight!

"In the morning the old chief harangued the men of his village, and ordered us all to get ready for the surround. About nine o'clock, every man war mounted, and we began to move.

"That war a sight to make a man's blood warm! A thousand men, all trained hunters, on horseback, carrying their guns, and with their horses painted in the height of Indians' fashion. We advanced until within about half a mile of the herd, then the chief ordered us to deploy to the right and left, until the wings of the column extended a long way, and advanced again.

"By this time the buffalo war all moving, and we had come to within a hundred yards of them. *Kow-e-so-te* then gave us the word, and away we went, pell-mell. Heavens, what a charge!

What a rushing and roaring—men shooting, buffalo bellowing and trampling until the earth shook under them!

"It war the work of half an hour to slay two thousand or may be three thousand animals. When the work was over, we took a view of the field. Here and there and everywhere, laid the slain buffalo. Occasionally, a horse with a broken leg war seen, or a man with a broken arm, or maybe he had fared worse, and had a broken head.

"Now came out the women of the village to help us butcher and pack up the meat. It war a big job, but we war not long about it. By night the camp war full of meat, and everybody merry. Bridger's camp, which war passing that way, traded with the village for fifteen hundred buffalo tongues—the tongue being reckoned a choice part of the animal. And that's the way we helped the Nez Perces hunt buffalo."

"But when you were hunting for your own subsistence in camp, you sometimes went out in small parties?"

"Oh yes, it war the same thing on a smaller scale. One time Kit Carson and myself, and a little Frenchman, named Marteau, went to run buffalo on Powder River. When we came in sight of the band it war agreed that Kit and the Frenchman should do the running, and I should stay with the pack animals. The weather war very cold and I didn't like my part of the duty much.

"The Frenchman's horse couldn't run, so I lent him mine. Kit rode his own, not a good buffalo horse either. In running, my horse fell with the Frenchman, and nearly killed him. Kit, who couldn't make his horse catch, jumped off, and caught mine, and tried it again. This time he came up with the band, and killed four fat cows.

"When I came up with the pack animals, I asked Kit how he came by my horse. He explained, and wanted to know if I had seen anything of Marteau: said my horse had fallen with him, and he thought killed him. 'You go over the other side of yon hill, and see,' said Kit.

"What'll I do with him if he is dead?" said I.

"Can't you pack him to camp?"

"Pack h—1" said I, "I should rather pack a load of meat."

"Waal," said Kit, "I'll butcher, if you'll go over and see, anyhow."

"So I went over, and found the dead man leaning his head on his hand, and groaning, for he war pretty bad hurt. I got him on his horse, though, after a while, and took him back to whar Kit war at work. We soon finished the butchering job, and started back to camp with our wounded Frenchman, and three loads of fat meat."

"You were not very compassionate toward each other, in the mountains?"

"That war not our business. We had no time for such things. Besides, live men war what we wanted, dead ones war of no account."

# 20

## 1838

From Missouri Lake, Meek started alone for the Gallatin Fork of the Missouri, trapping in a mountain basin called Gardiner's Hole. Beaver were plenty here, but it was getting late in the season, and the weather was cold in the mountains. On his return, in another basin called the Burned Hole, he found a buffalo skull, and knowing that Bridger's camp would soon pass that way, wrote on it the number of beaver he had taken, and also his intention to go to Fort Hall to sell them.

In a few days, the camp passing found the skull, which grinned its threat at the angry Booshways, as the chuckling trapper had calculated that it would. To prevent its execution runners were sent after him, who, however, failed to find him, and nothing was known of the supposed renegade for some time. But as Bridger passed through Pierre's Hole, on his way to Green River to winter, he was surprised at Meek's appearance in camp. He was soon invited to the lodge of the Booshways, and called to account for his supposed apostacy.

Meek, for a time, would neither deny nor confess, but put on his free trapper airs, and laughed in the face of the Booshways. Bridger, who half suspected some trick, took the matter lightly, but Dripps was very much annoyed, and made some threats, at which Meek only laughed the more. Finally, the certificate from their own trader, Jo Walker, was produced, the new pack of furs surrendered, and Dripps' wrath turned into smiles of approval.

Here again, Meek parted company with the main camp, and went on an expedition with seven other trappers, under John Larison, to the Salmon River: but found the cold very severe on this journey, and the grass scarce and poor, so that the company lost most of their horses.

On arriving at the Nez Percé village in the Forks of the Salmon, Meek found the old chief *Kow-e-so-te* full of the story of the missionaries and their religion, and anxious to hear preaching. Reports were continually arriving by the Indians, of the wonderful things which were being taught by Mr. and Mrs. Spalding at Lapwai, on the Clearwater, and at Waiilatpu, on the Walla-Walla River. It was now nearly two years since these missions had been founded, and the number of converts among the Nez Perces and Flatheads was already considerable.

Here, was an opening for a theological student, such as Joe Meek was! After some little assumption of modesty, Meek intimated that he thought himself capable of giving instruction on religious subjects, and being pressed by the chief, finally consented to preach to *Kow-e-so-te's* people. Taking care first to hold a private council with his associates, and binding them not to betray him, Meek preached his first sermon that evening, going regularly through with the ordinary services of a *meeting*.

These services were repeated whenever the Indians seemed to desire it, until Christmas. Then, the village being about to start upon a hunt, the preacher took occasion to intimate to the chief that a wife would be an agreeable present. To this, however, *Kow-e-so-te* demurred, saying that Spalding's religion did not permit men to have two wives: that the Nez Percé had many of them given up their wives on this account, and that, therefore, since Meek already had one wife among the Nez Percé , he could not have another without being false to the religion he professed.

To this perfectly clear argument Meek replied, that among white men, if a man's wife left him without his consent, as his had done, he could procure a divorce, and take another wife. Besides, he could tell him how the Bible related many stories of its best men having several wives. But *Kow-e-so-te* was not easily convinced. He could not see how, if the Bible approved of polygamy, Spalding should insist on the Indians putting away all but one of their wives. "However," says Meek, "after about two weeks' explanation of the doings of Solomon and David, I succeeded in getting the chief to give me a young girl, whom I called Virginia—my present wife, and the mother of seven children."

After accompanying the Indians on their hunt to the Beaverhead country, where they found plenty of buffalo, Meek remained with the Nez Percé village until about the first of March, when he again intimated to the chief that it was the custom of white men to pay their preachers. Accordingly, the people were notified, and the winter's salary began to arrive. It amounted altogether to thirteen horses, and many packs of beaver, beside sheep-skins and buffalo robes, so that he *considered that with his young wife, he had made a pretty good winter's work of it.*

In March he set out trapping again, in company with one of his comrades named Allen, a man to whom he was much attached. They traveled along up and down the Salmon, to Godin's River, Henry's Fork of the Snake, to Pierre's Fork, and Lewis's Fork, and the Muddy, and finally set their traps on a little stream that runs out of the pass which leads to Pierre's Hole.

Leaving their camp one morning to take up their traps, they were discovered and attacked by a party of Blackfeet just as they came near the trapping ground. The only refuge at hand was a thicket of willows on the opposite side of the creek, and toward this, the trappers directed their flight. Meek, who was in advance, succeeded in gaining the thicket without being seen, but Allen stumbled and fell in crossing the stream, and wet his gun. He quickly recovered his footing and crossed over, but the Blackfeet had seen him enter the thicket, and came up to within a short distance, yet not approaching too near the place where they knew he was concealed. Unfortunately, Allen, in his anxiety to be ready for defense, commenced snapping caps on his gun to dry it. The quick ears of the savages caught the sound, and understood the meaning of it. Knowing him to be defenseless, they plunged into the thicket after him, shooting him almost immediately, and dragging him out, still breathing to a small prairie about two rods away.

And now commenced a scene which Meek was compelled to witness, and which he declares nearly made him insane through sympathy, fear, horror, and suspense as to his own fate. Those devils incarnate deliberately cut up their still palpitating victim into a hundred pieces, each taking a piece, accompanying the horrible and inhuman butchery with every conceivable gesture of contempt for the victim, and of hellish delight in their own acts.

Meek, who was only concealed by the small patch of willows, and a pit in the sand hastily scooped out with his knife until it was deep enough to lie in, was in a state of the most fearful excitement. All day long he had to endure the horrors of his position. Every moment seemed an hour, every hour a day, until when night came, and the Indians left the place, he was in a high state of fever.

About nine o'clock that night he ventured to creep to the edge of the little prairie, where he lay and listened a long time, without hearing anything but the squirrels running over the dry leaves, but which he constantly feared was the stealthy approach of the enemy. At last, however, he summoned courage to crawl out on to the open ground, and gradually to work his way to a wooded bluff not far distant. The next day he found two of his horses, and with these set out alone for Green River, where the American Company was to rendezvous. After twenty-six days of solitary and cautious travel, he reached the appointed place in safety, having suffered fearfully from the recollection of the tragic scene he had witnessed in the death of his friend, and also from solitude and want of food.

The rendezvous of this year was at Bonneville's old fort on Green River[1], and was the last one held in the mountains by the

---

1. Since Joe reports meeting the Farnham party later this year, this was the rendezvous of 1839, near Horse Creek on the Green, not the one of 1840, which

American Fur Company. Beaver was growing scarce, and competition was strong. On the disbanding of the company, some went to Santa Fe, some to California, others to the Lower Columbia, and a few remained in the mountains trapping, and selling their furs to the Hudson's Bay Company at Fort Hall. As to the leaders, some of them continued for a few years longer to trade with the Indians, and others returned to the States, to lose their fortunes more easily far than they made them.

Of the men who remained in the mountains trapping, that year, Meek was one. Leaving his wife at Fort Hall, he set out in company with a Shawnee, named Big Jim, to take beaver on Salt River, a tributary of the Snake. The two trappers had each his riding and his packhorse, and at night generally picketed them all, but one night Big Jim allowed one of his to remain loose to graze. This horse, after eating for some hours, came back and laid down behind the other horses, and every now and then raised up his head, which slight movement at length aroused Big Jim's attention, and his suspicions also.

"My friend," said he in a whisper to Meek, "Indian steal our horses."

"Jump up and shoot," was the brief answer.

Jim shot, and ran out to see the result. Directly he came back saying, "My friend, I shoot my horse, break him neck." And Big Jim became disconsolate over what his white comrade considered a very good joke.

---

was also on the Green. In that case this was not quite the last held by American Fur: in fact the caravan brought west by Drips, Bridger, and Fraeb the next year was much larger, swelled by a number of missionaries. Though commonly called American Fur, the suppliers of rendezvous in 1839 and 1840 were successors of that firm, which was no longer in business—Hafen series, I, p. 162-164.

The hunt was short and not very remunerative in furs. Meek soon returned to Fort Hall, and when he did so, found his new wife had left that post in company with a party under Newell, to go to Fort Crockett, on Green River—Newell's wife being a sister of Virginia's—on learning which he started on again alone, to join that party. On Bear River, he fell in with a portion of that quixotic band, under Farnham, which was looking for paradise and perfection, something on the Fourier plan, somewhere in this western wilderness. They had already made the discovery in crossing the continent, that perfect disinterestedness was lacking among themselves, and that the nearer they got to their western paradise, the farther off it seemed in their own minds.[2]

Continuing his journey alone, soon after parting from Farnham, he lost the hammer of his gun, which accident deprived him of the means of subsisting himself, and he had no dried meat, nor provisions of any kind. The weather, too, was very cold, increasing the necessity for food to support animal heat. However, the deprivation of food was one of the accidents to which mountain men were constantly liable, and one from which he had often suffered severely, therefore he pushed on, without feeling any unusual alarm, and had arrived within fifteen miles of the fort before he yielded to the feeling of exhaustion, and laid down beside the trail to rest. Whether he would ever have finished the journey alone he could not tell, but fortunately for him, he was discovered by Jo Walker, and

2. Thomas Jefferson Farnham and his wife Eliza had been inspired by a lecture praising Oregon given by Jason Lee. The early West seems to have attracted quite a few utopians. One could be vastly amused at their attempts to set up model societies in the wilderness had not one such group—the one founded by Joseph Smith—succeeded so spectacularly.

Gordon, another acquaintance, who chanced to pass that way toward the fort.

Meek answered their hail, and inquired if they had anything to eat. Walker replied in the affirmative, and getting down from his horse, produced some dried buffalo meat which he gave to the famishing trapper. But seeing the ravenous manner in which he began to eat, Walker inquired how long it had been since he had eaten anything.

"Five days since I had a bite."

"Then, my man, you can't have any more just now," said Walker, seizing the meat in alarm lest Meek should kill himself.

"It was hard to see that meat packed away again," says Meek in relating his sufferings, "I told Walker that if my gun had a hammer, I'd shoot and eat him. But he talked very kindly, and helped me on my horse, and we all went on to the Fort."

At Fort Crockett were Newell and his party, the remainder of Farnham's party, a trading party under St. Clair, who owned the fort, Kit Carson, and a number of Meek's former associates, including Craig and Wilkins. Most of these men, Othello-like, had lost their occupation since the disbanding of the American Fur Company, and were much at a loss concerning the future. It was agreed betwen Newell and Meek to take what beaver they had to Fort Hall, to trade for goods, and return to Fort Crockett, where they would commence business on their own account with the Indians.

Accordingly they set out, with one other man belonging to Farnham's former adherents. They traveled to Henry's Fork, to Black Fork, where Fort Bridger now is, to Bear River, to Soda Springs, and finally to Fort Hall, suffering much from cold, and finding very little to eat by the way. At Fort Hall, which was still in charge of Courtenay Walker, Meek and Newell remained a week, when, having purchased their goods and horses to pack them, they once more set out on the long, cold journey to Fort Crockett. They had fifteen horses to take care of and only one assistant, a Snake Indian called Al. The return proved an arduous and difficult undertaking. The cold was very severe. They had not been able to lay in a sufficient stock of provisions at Fort Hall, and game there was none, on the route. By the time they arrived at Ham's Fork the only atom of food they had left was a small piece of bacon which they had been carefully saving to eat with any poor meat they might chance to find.

The next morning after camping on Ham's Fork was stormy and cold, the snow filling the air, yet Snake Al, with a promptitude by no means characteristic of him, rose early and went out to look after the horses.

"By that same token," said Meek to Newell, "Al has eaten the bacon." And so it proved, on investigation. Al's uneasy conscience having acted as a goad to stir him up to begin his duties in season. On finding his conjecture confirmed, Meek declared his intention, should no game be found before next day night, of killing and eating Al, to get back the stolen bacon. But Providence interfered to save Al's bacon. On the following afternoon, the little party fell in with another still smaller but better-supplied party of travelers, comprising a Frenchman and his wife. These had plenty of fat antelope meat, which they freely parted with to the needy ones, whom also they accompanied to Fort Crockett.

It was now Christmas, and the festivities which took place at the Fort were attended with a good deal of rum drinking, in which Meek, according to his custom, joined, and as a considerable portion of their stock in trade consisted of this article, it may fairly be presumed that the home consumption of these two *lone traders* amounted to the larger half of what they had with so much trouble transported from Fort Hall. In fact, *times were bad enough* among the men so suddenly thrown upon their own resources among the mountains, at a time when that little creature, which had made mountain life tolerable, or possible, was fast being exterminated.

To make matters more serious, some of the worst of the now unemployed trappers had taken to a life of thieving and mischief which made enemies of the friendly Indians, and was likely to prevent the better disposed from enjoying security among any of the tribes. A party of these renegades, under a man named Thompson, went over to Snake River to steal horses from the Nez Perces. Not succeeding in this, they robbed the Snake Indians of about forty animals, and ran them off to the Uintee, the Indians following and complaining to the whites at Fort Crockett that their people had been robbed by white trappers, and demanding restitution.

According to Indian law, when one of a tribe offends, the whole tribe is responsible. Therefore, if whites stole their horses, they might take vengeance on any whites they met, unless the property was restored. In compliance with this well-understood requisition of Indian law, a party was made up at Fort Crockett to go and retake the horses, and restore them to their rightful owners. This party consisted of Meek, Craig, Newell, Carson, and twenty-five others, under the command of Jo Walker.

The horses were found on an island in Green River, the robbers having domiciled themselves in an old fort at the mouth of the Uintee. In order to avoid having a fight with the renegades, whose white blood the trappers were not anxious to spill, Walker made an effort to get the horses off the island undiscovered. But while horses and men were crossing the river on the ice, the ice sinking with them until the water was knee-deep, the robbers discovered the escape of their booty, and charging on the trappers tried to recover the horses. In this effort they were not successful, while Walker made a masterly flank movement and getting in Thompson's rear, ran the horses into the fort, where he stationed his men, and succeeded in keeping the robbers on the outside. Thompson then commenced giving the horses away to a village of Utes in the neighborhood of the fort, on condition that they should assist in retaking them. On his side, Walker threatened the Utes with dire vengeance if they dared interfere. The Utes who had a wholesome fear not only of the trappers, but of their foes the Snakes, declined to enter into the quarrel. After a day of strategy, and of threats alternated with arguments, strengthened by a warlike display, the trappers marched out of the fort before the faces of the discomfited thieves, taking their booty with them, which was duly restored to the Snakes on their return to Fort Crockett, and peace secured once more with that people.[3]

Still, times continued bad. The men not knowing what else to do, went out in small parties in all directions seeking adventures, which generally were not far to find. On one of these excursions Meek went with a party down the canyon of Green River, on the ice. For nearly a hundred miles they traveled down this awful

3. For a fuller account of this incident of the recovery of the Indians' stolen horses, see Hafen, LeRoy R., and Ann W. Hafen, To the Rockies and Oregon, Glendale, California, 1955.

canyon without finding but one place where they could have come out, and left it at last at the mouth of the Uintee.

This passed the time until March. Then, the company of Newell and Meek was joined by Antoine Rubideau, who had brought goods from Sante Fe to trade with the Indians. Setting out in company, they traded along up Green River to the mouth of Ham's Fork and camped. The snow was still deep in the mountains, and the trappers found great sport in running antelope. On one occasion a large herd, numbering several hundreds, were run on to the ice, on Green River, where they were crowded into an air hole, and large numbers slaughtered only for the cruel sport which they afforded.

But killing antelope needlessly was not by any means the worst of amusements practiced in Rubideau's camp. That foolish trader occupied himself so often and so long in playing *Hand*—an Indian game—that before he parted with his new associates, he had gambled away his goods, his horses, and even his wife, so that he returned to Santa Fe much poorer than nothing—since he was in debt.

On the departure of Rubideau, Meek went to Fort Hall, and remained in that neighborhood, trapping and trading for the Hudson's Bay Company, until about the last of June, when he started for the old rendezvous places of the American companies, hoping to find some divisions of them at least, on the familiar camping ground. But his journey was in vain.[4] Neither on Green River or Wind River, where for ten years he had been

---

4. Though Meek says he was unable to find his old comrades holding a rendezvous this summer, one was held on Green River starting June 30—Hafen series, I, p. 163. Joe's friend, brother-in-law, and usual trapping companion of these days, Doc Newell, did get to the rendezvous and found it dull—Newell, p.

accustomed to meeting the leaders and their men, his old comrades in danger, did he find a wandering brigade even. The glory of the American companies was departed, and he found himself solitary among his long, familiar haunts.

With many melancholy reflections, the man of twenty-eight years of age recalled how, a mere boy, he had fallen half unawares into the kind of life he had ever since led among the mountains, with only other men equally the victims of circumstance, and the degraded savages, for his companions. The best that could be made of it, such life had been and must be constantly deteriorating to the minds and souls of himself and his associates. Away from all laws, and refined habits of living, away from the society of religious, modest, and accomplished women, always surrounded by savage scenes, and forced to cultivate a taste for barbarous things—what had this life made of him? What was he to do with himself in the future?

Sick of trapping and hunting, with brief intervals of carousing, he felt himself to be. And then, even if he were not, the trade was no longer profitable enough to support him. What could he do? Where could he go? He remembered his talk with Mrs. Whitman, that fair, tall, courteous, and dignified lady who had stirred in him longings to return to the civilized life of his native state. But he felt unfit for the society of such as she. Would he ever, could he ever attain to it now? He had promised her he might go over into Oregon and settle down. But could he settle down? Should he not starve at trying to do what other men, mechanics and farmers do? And as to learning, he had

---

39. And despite Mrs. Victor's assertion that he was 28, Meek was 30 years old at this time.

none of it. There was no hope then of *living by his wits*, as some men did—missionaries and artists and school teachers, some of whom he had met at the rendezvous. Heigho! To be checkmated in life at twenty-eight, that would never do.

At Fort Hall, on his return, he met two more missionaries and their wives going to Oregon[5], but these four did not affect him pleasantly, he had no mind to go with them. Instead, he set out on what proved to be his last trapping expedition, with a Frenchman named Mattileau. They visited the old trapping grounds on Pierre's Fork, Lewis's Lake, Jackson's River, Jackson's Hole, Lewis River and Salt River, but beaver were scarce, and it was with a feeling of relief that, on returning by way of Bear River, Meek heard from a Frenchman whom he met there, that he was wanted at Fort Hall, by his friend Newell, who had something to propose to him.

---

5. Tobie Hafen series, I, p. 324 identifies the full party taking the wagons westward, aside from Meek and Newell, as Caleb Wilkins, William Craig, John Larison, a German called Nicholas, and a Shoshone Indian. The wife and baby Meek refers to were Virginia Meek and Courtney Walker Meek. Newell and Wilkins likewise had Nez Percé wives whom they took to settle in Oregon.

# 21

**1840**

When Meek arrived at Fort Hall, where Newell was awaiting him, he found that the latter had there the two wagons which Dr. Whitman had left at the points on the journey where further transportation by their means had been pronounced impossible. The Doctor's idea of finding a passable wagon-road over the lava plains and the heavily timbered mountains lying between Fort Hall and the Columbia River, seemed to Newell not so wild a one as it was generally pronounced to be in the mountains. At all events, he was prepared to undertake the journey. The wagons were put in traveling order, and horses and mules purchased for the expedition.

"Come," said Newell to Meek, "we are done with this life in the mountains—done with wading in beaver-dams, and freezing or starving alternately—done with Indian trading and Indian fighting. The fur trade is dead in the Rocky Mountains, and it is no place for us now, if ever it was. We are young yet, and have life before us. We cannot waste it here. We cannot or will not return to the States. Let us go down to the Wallamet and take farms. There is already quite a settlement there made by the Methodist Mission and the Hudson's Bay Company's retired servants.

"I have had some talk with the Americans who have gone down there, and the talk is that the country is going to be settled up by our people, and that the Hudson's Bay Company are not going to rule this country much longer. What do you say, Meek? Shall we turn American settlers?"

"I'll go where you do, Newell. What suits you suits me."

"I thought you'd say so, and that's why I sent for you, Meek. In my way of thinking, a white man is a little better than a Canadian Frenchman. I'll be damned if I'll hang 'round a post of the Hudson's Bay Company. So you'll go?"

"I reckon I will! What have you got for me to do? *I* haven't got anything to begin with but a wife and baby!"

"Well, you can drive one of the wagons, and take your family and traps along. Nicholas will drive the other, and I'll play leader, and look after the train. Craig will go also, so we shall be quite a party, with what strays we shall be sure to pick up."

Thus it was settled. Thus Oregon began to receive her first real emigrants, who were neither fur-traders nor missionaries, but true frontiersmen—border-men. The training which the mountain men had received in the service of the fur companies admirably fitted them to be, what afterward they became, a valuable and indispensable element in the society of that country in whose peculiar history they played an important part. But we must not anticipate their acts before we have witnessed their gradual transformation from lawless rangers of the wilderness to law-abiding and even law-making and law-executing citizens of an isolated territory.

# A Look at Book Two:
## The River of the West, The Oregon Years

***Filled with raw adventure, the joyous story of life in the untamed wilds of the American frontier continues.***

In this second volume, Joe Meek comes to light as pioneer, sheriff, US Marshal, and legislator. Through his wondrous adventures, he becomes an important part of Oregon's formative years, as well as that of the entire Northwest.

Authentically human, Meek details the struggles of women, Native Americans, missionaries, trappers, early settlers, explorers, and the Hudson's Bay company, and how they all came together to shape a territory—and finally a state.

*With vivid descriptions of the wilderness and Meek's unfiltered observations, Win Blevins brings to life a bygone era of adventure and danger in this colorful memoir.*

***AVAILABLE AUGUST 2024***

# Acknowledgments

Thanks go to several scholars who generously took the time to answer persnickety questions: E. L. Bulow of Gallup, New Mexico, George Miles, curator of the Western Americana collection at Yale's Beinecke Library, Professor George Belknap of the University of Oregon, and especially Hazel Mills of Olympia, Washington.

Thanks as well to Michael Kelly of the Buffalo Bill Historical Center in Cody, Wyoming, and to the patient staff of Hot Springs County Library in Thermopolis, Wyoming.

# Further Reading On Joe Meek

Neither of the two book-length treatments of Meek's life may equal his own telling for color and style, but each has its value:

Harvey E. Tobie's *No Man Like Joe*, unfortunately out of print, is scholarly and reliable. So is his biographical essay about Meek in LeRoy Hafen's *The Mountain Men and the Fur Trade of the Far West*, Volume I, p. 313-335. Recommended as antidote to some of Meek's tall tales and his getting confused about what year it was.

Stanley Vestal's biography *Joe Meek: The Merry Mountain Man*, though it sometimes seems boys' adventure stuff, catches Joe's flair. Out of print in cloth, it is available in paperback.

Checking the tale as told by someone else who was there can be illuminating, and a generous handful of books—written by them or someone who spoke with them directly—lets us do that: Osborne Russell's *Journal of a Trapper*, Kit Carson's *Autobiography*, T. D. Bonner's *The Life and Adventures of James P. Beckwourth*, Warren Ferris' *Life in the Rocky Mountains*, Zenas

Leonard's *Adventures*, George Frederick Ruxton's *Life in the Far West*, William Drummond Stewart's *Edward Warren*, Washington Irving's *The Adventures of Captain Bonneville*, among others. These two by Meek's trapper brother and his long-time trapping companion, respectively, are of interest for detail: Stephen Hall Meek's *The Autobiography of a Mountain Man*, and Robert Newell's *Memoranda of Travel in Missouri*.

The biographies of some of Meek's noted contemporaries fill out our picture of Joe and the time and the people: See especially Dale Morgan's *Jedediah Smith and the Opening of the West*. Cecil Alter's *Jim Bridger*, John Sunder's *Bill Sublette, Mountain Man*, Elinor Wilson's *Jim Beckwourth*, Alpheus Favour's *Old Bill Williams*, Hafen's *Broken Hand: The Life of Thomas Fitzpatrick* Harvey L. Carter's *Dear Old Kit: the Historical Christopher Carson*, and M. Morgan Estergreen's *Kit Carson: A Portrait in Courage*.

*Valuable one-volume histories are Bernard DeVoto's Across the Wild Missouri, Don Berry's A Majority of Scoundrels: an informal history of the Rocky Mountain Fur Company, and Winfred Blevins's Give Your Heart to the Hawks: A Tribute to the Mountain Men.*

# About the Author

Win Blevins was an award-winning author best known for his fiction and non-fiction books of Western lore and Native American leaders, lifestyle, and spirituality. He was the recipient of a lifetime achievement award from the Western Writers of America, and a member of the Western Writers Hall of Fame; a three-time winner of Wordcraft Circle Native Writers and Storytellers Book of the Year; two-time winner of a Spur Award for Best Novel of the West; and was nominated for a Pulitzer for his novel about Crazy Horse, *Stone Song*.

Blevins, whose own origins were a mix of Cherokee, Welsh-Irish, and African American, published his first novel in 1973. That book, *Give Your Heart to the Hawks, a Tribute to the Mountain Man*, is still in print fifty years later and recently returned to the *New York Times* bestseller list.

Over his long career, Blevins wrote nearly forty books, including the historical fiction Rendezvous series, a dozen screenplays, and numerous magazine articles. His *Dictionary of the American West* is held in 750 libraries.

Born in Little Rock, Arkansas, on October 21, 1938, Blevins was an honors graduate of Columbia University—where he earned a master's degree—and the Music Conservatory of the University of Southern California. He began his writing career as a music and drama critic for the *Los Angeles Times* and became the principal entertainment editor for the *The Los Angeles Herald Examiner*. During that time, he hung out with the likes of Sam Peckinpah and Strother Martin, and began diving into the lives of Mountain Men and Native Americans of the West.

He also served as the Gaylord Family Visiting Professor of Professional Writing at the University of Oklahoma. For fifteen years, he was a book editor for Macmillan Publishing and TOR/Forge Books.

Win loved and felt a deep connection with nature. He climbed mountains on four continents and was a boatman-guide on the Snake River. Once caught in a freak blizzard while climbing, he took shelter inside a tree for more than twenty-four hours. His feet were frozen, but he refused to have them amputated. Almost twenty years after that event, he climbed the Himalayas—despite an awkward gait.

Native Spirituality suited him. He was pierced during a Lakota ceremony and was a pipe carrier. He went on twelve vision quests and felt the pull of the red road.

Win spent the last twenty years of his life, living quietly in the Southwest among the Navajo. His passions grew with time. In the center was his wife Meredith, their children, and many grandchildren. Classical music, baseball, roaming red rock mesas, and rafting were great loves, and he considered himself blessed to create new stories about the West. He was also proud to call himself a member of the world's oldest profession—storytelling.

Printed in the USA
CPSIA information can be obtained
at www.ICGtesting.com
CBHW011511250624
10644CB00012B/123

9 781639 775200